14

APPUNTI

LECTURE NOTES

Renata Scognamillo and Umberto Zannier
Scuola Normale Superiore
Piazza dei Cavalieri, 7
56126 Pisa

Introductory Notes on Valuation Rings and Function Fields in One Variable

Renata Scognamillo
and Umberto Zannier

Introductory Notes on Valuation Rings and Function Fields in One Variable

EDIZIONI
DELLA
NORMALE

ISBN 978-88-7642-500-4
ISBN 978-88-7642-501-1 (eBook)

Contents

Preface

The present notes arise from courses delivered at the Scuola Normale in the years 2004/2006. These courses were intended to give to students of III and IV at Scuola Normale year some background on function fields of one variable, from the viewpoint of valuations, these topics not being usually touched in university courses in Pisa. The initial purpose was to limit the treatment to basic theory.

The notes follow very similar principles, and are addressed to a public with similar prerequisites.

More precisely, the main part consists of two chapters, after a brief introduction. The first one introduces very basic notions of algebraic geometry, especially concerning function fields of algebraic curves. This is intended mainly as a motivation for the second chapter, devoted to the study of valuation rings and their completions. (This follows more or less standard approaches; however we have added occasionally some less standard material, and we hope that this could be not entirely free of interest even for more expert readers.)

The notes contain several exercises, and are concluded with three appendices. The first one discusses Hilbert's Nullstellensatz, for which a number of proofs are presented. The second one discusses Puiseux series; they are introduced in the second chapter, but here certain arithmetical aspects are considered in more detail. The third appendix instead contains some factorisation theory in Dedekind domains, applying tools from Chapter 2.

In subsequent courses this material was expanded, to cover basic geometry of curves, including for instance the theorem of Riemann-Roch. It was our intention to include all of this in an expanded version of the notes, but the last part is still in incomplete shape, and then we decided to publish only the present ones. We hope we shall be able to publish such a more complete version in a not too far future.

The prerequisites for this course are rather small: usually a very basic knowledge of field theory and commutative algebra is sufficient. (More specifically, we assume the elementary theory of algebraic and transcendental extensions of fields, the notion of Noetherian ring and Hilbert's basis theorem, local rings and Nakayama's lemma and basic properties of polynomials. References for the needed results are always indicated.)

We have borrowed freely from a number of sources (quoted throughout) but we have added some material and combined a few things which commonly do not appear together in a same presentation.

Introduction
Generalities on algebraic functions of one variabile

We shall consider *algebraic functions of one variable*. They appear naturally, for instance in elementary Analysis (real or complex), just after rational functions; think for instance of radicals $\sqrt[d]{r(x)}$ of rational functions $r(x)$. More generally, roughly speaking we can think of functions $\varphi(x)$, say of the real or complex variable x, such that $F(x, \varphi(x)) = 0$ for the values of x in question and for some nonzero polynomial $F(X, Y)$.

Such a viewpoint is legitimate (and unavoidable), however it soon leads to the necessity to add several and fundamental precisions. For instance:

- Which is the natural domain of an algebraic function? (The function and domain may be already given, but one could ask for an extension of the given domain, and of the function, required to satisfy some kind of regularity.)
- Which type of regularity can we expect for such functions?

And moreover, we wish, reciprocally, to construct algebraic functions starting only from the knowledge of the algebraic equation, *i.e.*, we wish that a certain polynomial $F(X, Y)$ as above would define (one or several) algebraic functions, as solutions of $F(x, \varphi(x)) = 0$.

However, for a given x, there are in general $\deg_Y F$ possibilities for the value $\varphi(x)$; and hence, how to choose among them? And, even more important, how to coordinate these choices for varying x?

So, in the first place, in order to guarantee the existence of solutions it shall be desirable to work, for the moment with the complex field as a ground field.

With this in mind, for a given polynomial F, supposed for simplicity to be monic in Y, we could then agree to choose a solution y_0 of $F(x_0, Y) = 0$, at a given point $x_0 \in \mathbb{C}$, and to construct a function $\varphi(x)$ such that $\varphi(x_0) = y_0$, $F(x, \varphi(x)) = 0$ and such that φ is 'sufficiently' regular, *e.g.*, continuous.

The Implicit Function theorem (together with some topology) ensures a (unique) continuous extension of φ to a domain $\mathcal{D} \subset \mathbb{C}$, simply connected and with $x_0 \in \mathcal{D}$, *with the condition that $\mathcal{D}$ does not contain any point x_1 such that the polynomial $F(x_1, Y)$, in the variable Y, has multiple roots*. In fact, let us observe that with such assumption the set $\Gamma := \{(x, y) \in \mathbb{C}^2 | x \in \mathcal{D}, F(x, y) = 0\}$ is (by the Implicit function theorem and by the assumption on $\mathcal{D}$) a covering space of $\mathcal{D}$ (with respect to the first projection - recall that F is supposed here to be monic in Y) and the assertion then follows from the Monodromy Theorem (see, *e.g.*, [11] or the more introductory presentation in [13]).

The assumption about multiple roots is in general necessary in order to extend our solution $\varphi(x)$ to a neighborhood of x_1: think for instance at the simple case of $\sqrt{x}$ (where $F(X, Y) = Y^2 - X$); this may be defined as a continuous function *e.g.* in the domain $\{x \in \mathbb{C}, 0 < |x| < 1, x \notin (0, 1)\}$, however without any possibility of extending it to a continuous function on a whole neighborhood of 0. (See next exercise.)

Exercise 1. Letting $D \subset \mathbb{C}$ be any open disk centered at 0, prove in a direct way that there is no continuous function $\rho : D \to \mathbb{C}$ such that $\rho(x)^2 = x$ for all $x \in D$.
(Hint: first show that it suffices to prove the assertion with D replaced by the unit circle $S_1 := \{x \in \mathbb{C} : |x| = 1\}$ and a $\rho : S_1 \to S_1$. For this, either prove that a continuous $\rho(x)$ would send S_1 into an open and closed subset of S_1, and obtain a contradiction; or, alternatively, set, for real t, $\psi(t) := \rho(e^{2\pi i t})e^{-\pi i t}$. Then ψ would be a continuous function from $\mathbb{R}$ to S_1 such that $\psi^2(t) = 1$. Hence ψ would take values in $\{\pm 1\}$ and would hence be constant, say equal to $c \in \{\pm 1\}$. Then $\rho(e^{2\pi i t}) = ce^{\pi i t}$, which yields a contradiction on evaluating at $t = 0$ and $t = 1$.)

Therefore we encounter a serious difficulty in trying to construct, starting from F, a globally defined function on $\mathbb{C}$. The outlined procedure leads to a construction of *local sections of a cover* rather than *global* algebraic functions. The cover is obtained as above, on defining, given the polynomial F, $\mathcal{D}_F := \{x \in \mathbb{C} | F(x, Y) \text{ has no multiple roots}\}$ and $\Gamma_F = \{(x, y) \in \mathbb{C}^2 | x \in \mathcal{D}, F(x, y) = 0\}$. Since F is monic in Y, Γ_F is a (not necessarily connected) cover of $\mathcal{D}_F$, of degree $\deg_Y F$. Clearly, in general $\mathcal{D}_F$ shall not be simply connected, hence in this way we shall be able to obtain algebraic functions only locally, on suitable subsets.

Such a viewpoint, although not leading *tout court* to global algebraic functions, is anyway extremely important, and links the topic with fundamental topological concepts. The local sections so constructed turn out, by the way, to be automatically analytic.

1. A first viewpoint for algebraic functions: Riemann surfaces

It was Riemann the first to conceive algebraic functions as defined not quite on $\mathbb{C}$ (or subsets of it), but rather, roughly speaking, on the space V_F of solutions $(a, b) \in \mathbb{C}^2$ of $F(x, y) = 0$; this space constitutes the *plane affine complex curve* determined by $F(X, Y)$. What was the *variable*, *i.e.* x, becomes now merely a coordinate function, with a role not anymore special with respect to y.

Suppose now that F is irreducible (which is not a loss of generality). Then one may prove that V_F, once deprived of the *singular points* (those for which $F(a, b) = F_X(a, b) = F_Y(a, b) = 0$) becomes a real connected surface equipped with a complex structure (and therefore orientable), *i.e.* a *Riemann surface*.[1] If we suitably compactify it, adding a certain finite set of points, one may prove that the resulting space $\mathcal{R}$ is homeomorphic to a *torus with g handles* (where $g = g_{\mathcal{R}}$ is an integer ≥ 0). The coordinate functions x, y on the original space may be continued to regular holomorphic functions from $\mathcal{R}$ to $\mathbb{P}_1(\mathbb{C})$, related by the algebraic relation $F(x, y) = 0$. They generate a field $K := \mathbb{C}(x, y)$, called the field of rational functions of the Riemann surface in question. One proves that it coincides with the field of meromorphic functions for the underlying complex structure on $\mathcal{R}$.

Let us observe that $K = \mathbb{C}(x, y)$ has transcendence degree 1 over the *constant field* $\mathbb{C}$; the terminology 'of one variable' comes from this issue (and from the previous, related, discussion). (See also Definition 1.1.1 below.)

It is a fundamental fact that, up to biholomorphisms, $\mathcal{R}$ is determined not quite by the polynomial F or by the functions x, y which generate the field (over $\mathbb{C}$), but rather by the field K itself. Naturally such field can be generated by several other pairs (or sets) of elements in it. A special set of generators determined a 'model' of K; the generators determine a corresponding set of equations which define the algebraic relations among the generators themselves. The polynomials corresponding to such relations generate an ideal, which is in any case finitely generated by a basic theorem of Hilbert (see [2]). The space of common zeros to all polynomials in the ideal determines an *algebraic curve*, which constitutes a model of K in geometric sense.

For example, it is a simple fact to prove (see below) that in the case of two generators, denoted as above x, y, the ideal of relations is generated by a single irreducible polynomial $F(X, Y)$ (so that $F(x, y) = 0$). The

---

[1] For the theory, see [7, 11, 13, 16, 32] or [33]. Here we shall only recall a few facts.

space of the pairs (a, b) with $F(a, b) = 0$ then constitutes the geometric model of K corresponding to x, y.

2. A second viewpoint: geometry of curves

A related viewpoint, in which however the complex topology does not play a direct role, is the one purely algebro-geometric one of curves. In this approach, $\mathbb{C}$ may be replaced by other fields k, even of positive characteristic. For simplicity we shall assume in what follows that k is algebraically closed.

Now we do not start from the field K, but from an algebraic curve, of which K shall be the *function field*. Let then be given for instance a plane curve, defined as the set of points $(a, b) \in k^2$ with $F(a, b) = 0$ (where F is supposed to be an irreducible polynomial in $k[X, Y]$). The curve yields a function field defined as the fraction field of the quotient ring $k[X, Y]/(F(X, Y))$ of the polynomial ring $k[X, Y]$ modulo the ideal generated by $F(X, Y)$ (this quotient being a domain because F is irreducible and therefore generates a prime ideal). If $\deg_Y F > 0$, one can check (see next exercise) that such field is of the shape $K = k(x, y)$, where x is transcendental over k and where y is an element of an algebraic extension of $k(x)$, satisfying $F(x, y) = 0$. Given F, such a field is well defined up to isomorphisms over k.

Exercise 2. Prove that if $F(X, Y) \in k[X, Y]$ is irreducible over k and of positive degree in Y, then the fraction field of the domain $k[X, Y]/(F)$ is isomorphic to $k(X)[Y]/(F)$ over k.

Concerning the points (a, b) of the curve, it proves convenient, as in the case of Riemann surfaces, to introduce further concepts in order to obtain intrinsic descriptions, dependent on K but not on the *model* of K determined by the equation in question.

For this purpose, one can first *complete* the curve by adding *points at infinity*; this is analogue to the compactification in the complex topology, and may be obtained on considering the homogenized polynomial $\tilde{F}(X, Y, Z) := Z^{\deg F} F(\frac{X}{Z}, \frac{Y}{Z})$ (homogeneous of degree $\deg F$) and the solutions $(a, b, c) \in k^3 \setminus (0, 0, 0)$, considered as elements of $\mathbb{P}_2(k)$, *i.e.* up to proportionality.

Secondly, one can remove *singular points*. Recall that in the complex case that we have touched above we had first removed them and then compactified the resulting space by adding suitable new points. This was necessary because in a neighborhood of a singular point the curve is not a manifold (it is not even homeomorphic to an open subset of $\mathbb{C}$).

Algebraically, it is similarly possible to *desingularize* the curve, obtaining a *complete and nonsingular (or 'smooth') model* of it, which is found to amount to a projective (*i.e.*, embeddable in some projective space $\mathbb{P}_n$) algebraic curve, whose function field is isomorphic to the function field K of the original curve. We remark that it is not always possible to obtain this as a plane curve (namely, inside $\mathbb{P}_2$).

An advantage of such projective smooth model of K is that every rational function to another projective variety may be extended to a function everywhere regular. A consequence is (as in the case of Riemann surfaces) the uniqueness of the model, up to invertible rational regular functions.

Again we obtain something intrinsically linked to K. (We remark that, however, not all the models of K give a biregular correspondence with a smooth complete model, a phenomenon which itself is a motivation to the desingularisation.)

3. A third viewpoint: fields

A third viewpoint, the one adopted in these notes, consists in studying directly the function field K, forgetting about any geometric model, no matter if it comes from complex geometry of Riemann surfaces, or from algebraic geometry of curves.

The properties which we shall assume for the field K shall be the ones which characterize it as the function field of an algebraic curve; namely, we shall have to start with a ground field k (denominated the *field of constants*), which we shall usually assume to be algebraically closed, for the sake of simplicity.[2] Then we shall assume that K/k is finitely generated and of transcendence degree 1. (These properties, as illustrated above, are satisfied for the fields of rational functions on a curve.)

In such purely algebraic approach naturally one has to keep in mind the geometrical meaning of the function field in question, in order to avoid artificial notions. The issue arises of how to recover the notion of *point*, directly from the field, and without geometric models. This indeed can be done indirectly; one has first to recall that in any geometric model to each point P is associated a ring $\mathcal{O}_P \subset K$ made up of those elements of K which correspond to rational functions which are defined at P. Now, in the case when P is nonsingular, this ring has certain remarkable properties: for instance, its fraction field is K and it admits a unique nonzero prime ideal, which moreover is principal. This ideal cor-

---

[2] It should be noted that very interesting questions arise without this limitation.

responds geometrically to the set of functions which vanish at P, and the fact that it is principal allows one to define the *order* of vanishing at P (like for polynomials, or for analytic functions). Reversing these considerations, we shall then study these rings, denominated *discrete valuation rings* (DVR), rather than the points corresponding to them; actually, the ring themselves shall be called 'points'.

The DVR are in turn a special case of *valuation rings* (VR), which also do generally appear in geometry, not only of curves but also in higher dimensions; and they appear as well in other algebraic contexts.

Therefore we shall start with the study of these last rings; then we shall specialize to the case of DVR and to the fields that we have in mind. We shall anyway always keep in mind the geometric viewpoint, illustrating its correspondence with the purely algebraic definitions.

Moreover, we shall often discuss situations analogues to the geometric case of curves, which appear for instance in the study of the rational field $\mathbb{Q}$ and its finite extensions. Indeed, $\mathbb{Q}$ has properties similar to a field $k(x)$, similarly to the fact that $\mathbb{Z}$ is analogue to the polynomial rings $k[x]$.

In this direction, let us also recall that historically it has been rather natural to study equations $f(x, y) = 0$ as a next case with respect to equations $f(y) = 0$ in a single variable. In this last case a question is *how many solutions does the equation admit?* We know there at most $\deg f$. Of course, for algebraically closed fields k, there are infinitely many solutions in several variables; however geometry and topology have suggested other numerical invariants. We shall not touch this here, limiting ourselves to a few examples and to the study of VR; we plan to return to curves and to enlarge the content in a next edition.

In the present notes we shall often assume that k is algebraically closed. For simplicity, we shall often assume also that it has zero characteristic, although this is an important restriction, which for instance prevents some remarkable arithmetical applications.

Chapter 1
Basic notions on function fields of one variable

This chapter shall be devoted mainly to present examples and exercises, related to very basic introductory concepts in algebraic geometry. These shall not be strictly necessary for the theory of VR, presented in the next chapter; however they will serve the purpose of better understanding the geometrical motivations for the relevant abstract algebraic notions, as illustrated in the Introduction.

1.1. Function fields of one variable, rational fields

Definition 1.1.1. Let K/k be a field extension. We say that K *is a (n algebraic) function field of one variable over k* (abbreviated FF or FF/k) if K/k is finitely generated and of transcendence degree 1.

In other words, there exists $x \in K$, transcendental over k, and there exist $y_1, \ldots, y_n \in K$, all algebraic over $k(x)$, such that $K = k(x, y_1, \ldots, y_n)$.

Sometimes in the definition it is assumed that k is algebraically closed in K; this shall often be the case in our examples and results.

Example 1.1.1 (Rational fields). The simplest, though very important, case occurs when $K = k(x)$ with x transcendental over k, *i.e.* when K/k may be generated by a single element. Such a (function) field is called *rational* (over k).

We have a similar definition in any number of variables: a finitely generated extension K/k is called rational if there exist $x_1, \ldots, x_d \in K$ algebraically independent over k, such that $K = k(x_1, \ldots, x_d)$.

In several ways, the rational function fields in one variable are analogous to $\mathbb{Q}$ (and in the same way the function fields are analogous to number fields, *i.e.* finite extensions of $\mathbb{Q}$). For instance, once a generator x for K/k is given, we can see K as the fraction field of the ring $k[x]$, which is an euclidean ring, analogous to $\mathbb{Z}$.

Naturally, it may well happen that a field $K = k(z_1, \ldots, z_r)$ described by several generators z_i can be of the shape $k(x)$ for a suitable x: consider for the sake of very simple example the field $K = k(\sqrt{x^3}, \sqrt{x^5})$, which of course equals $k(\sqrt{x})$; see also l'Exercise 2.2 and Exercise 6(b) for other significant cases. When this happens, the generators $z_1, \ldots, z_r$ can be written as rational functions $z_i = Z_i(x)$, $Z_i \in k(X)$. Clearly these rational functions shall identically verify any algebraic relation over k valid among the z_i. We then speak of *parametrization* (in terms of the 'parameter' x) of such algebraic relations. This may be useful for instance to find solutions to the given algebraic relations with elements of a given field (letting for instance x vary over that field, and provided the Z_i have also coefficients in that field). Another important example, also historically relevant, occurs with the calculation of indefinite integrals, in case the involved functions become rational functions after a suitable substitution.[1]

Example 1.1.2 (Generators for a function field). Although, as we shall see, not every FF/k may be generated by a single element (*i.e.*, there are non-rational function fields), in general we have that *every function field in one variable may be generated by two elements*. Let us see why, supposing for simplicity that char$(k) = 0$. It is then well known in the elementary theory of algebraic field extensions (see for instance [17]) that, writing as above $K = k(x, y_1, \ldots, y_n)$, there exists a *primitive element* y for the algebraic extension $K/k(x)$, namely such that $K = k(x, y)$ (and actually y may be expressed as a suitable linear combination over k of $y_1, \ldots, y_n$). One can also prove ([17]) that the assumption char$(k) = 0$ can be removed by a suitable choice of x (namely with $K/k(x)$ separable).

Example 1.1.3 (Automorphisms of rational function fields).   Let us consider again a rational function field K/k, so there exists x such that $K = k(x)$. Such an x shall not be unique: indeed, we have $K = k(y)$ for every $y = (ax + b)/(cx + d)$ where a, b, c, d are in k and $ad \neq bc$. It is very easy to check that the transformations $x \mapsto y$ of this shape form a group: indeed, the shape of these transformations may be seen as a way of writing the linear automorphisms of the projective line over k.

We note that, since $k(x) = k(y)$ (with y as above) every such transformation g yields an automorphism $\sigma = \sigma_g$ of $k(x)/k$ (i.e. an automorphism of $k(x)$ fixing k pointwise), through the formula $\sigma^{-1}(s(x)) := s(g(x)) = s(y)$ for an $s \in k(x)$.[2]

If we substitute an element of k in place of x, or rather an element of $\mathbb{P}_1(k) = k^2/k^* \cong k \cup \infty$, we obtain an automorphism of $\mathbb{P}_1(k)$ by means

---

[1] Indeed, indefinite integrals $\int R(t)dt$ of rational functions $R(t)$ may be expressed as rational functions plus a linear combination of functions of the shape $\log(t + a)$.

[2] The exponent '-1' appears to guarantee that $\sigma_{gh} = \sigma_g \circ \sigma_h$.

of the formula $(t : u) \mapsto (at + bu : ct + du)$ (here $\infty := (1 : 0)$); in this interpretation the group is denoted $PGL_2(k)$.

We shall soon show (after the next exercises) that all the automorphisms of $k(x)/k$ are of this shape.

Exercise 1.1.1.

(i) Prove that the action of $PGL_2(k)$ on $\mathbb{P}_1(k)$ is triply transitive, *i.e.* given two ordered triples (a_1, a_2, a_3) and (b_1, b_2, b_3) of distinct elements of $\mathbb{P}_1(k)$ (*i.e.*, the a_i are distinct and also the b_i are distinct), there exists a $g \in PGL_2(k)$ such that $g(a_i) = b_i$ for $i = 1, 2, 3$. prove also that such a g is unique (for this it will suffice, and shall simplify things, to study the stabilizer of $(0, 1, \infty)$).

(ii) Prove that the said action of $PGL_2(k)$ leaves invariant the 'cross-ratio' of four distinct points $a, b, c, d \in \mathbb{P}_1(k)$, namely the quantity $(a - c)(b - d)/(a - d)(b - c)$ (with obvious definitions when $\{a, b, c, d\}$ contains ∞). Prove moreover that $PGL_2(k)$ is transitive when acting on the ordered 4-tuples with a given cross-ratio.
(Hint: reduce to the case when three among a, b, c, d are $0, 1, \infty$.)

(iii) Write down explicitly the elements of the group which permute $(0, 1, \infty)$. They make up a subgroup isomorphic to S_3. Find a rational function $\lambda(x)$ of degree 6 invariant for this action and prove that the field of invariants inside $k(x)$ is just $k(\lambda(x))$.
(Hint: a little knowledge of Galois theory shall help.)

(iv) Prove that, given any four distinct points $a, b, c, d \in \mathbb{P}_1$ there exists $g \in PGL_2$ sending them onto $l, -l, 1/l, -1/l$ for some l.

(v) For $P \in \mathbb{P}_1(k)$, let $G = G_P$ be the subgroup of $PGL_2(k)$ fixing P. Prove that if $\mathrm{char}(k) = 0$ every finite subgroup of G is cyclic.
(Hint: reduce to the case $P = \infty$.)
Prove that every element of $PGL_2(k)$ has one or two fixed points. Prove that in the first case the element is conjugate to a translation $t \mapsto t + c$, whereas in the second case it is conjugate to a map $t \mapsto ct$.
(Hint: conjugate by a transformation sending the fixed point(s) to ∞ or $\infty, 0$.)

(vi) Prove that for $d \geq 3$ the group $PGL_2(k)$ does not contain any subgroup isomorphic to $\mathbb{Z}/(d) \times \mathbb{Z}/(d)$, but that it does for $d = 2$. More generally, describe the finite abelian groups contained in $PGL_2(k)$.[3]

---

[3] The structure of finite subgroups of $PGL_2(k)$ is well known; for instance in characteristic zero they can only be cyclic, dihedral, or isomorphic to S_3, S_4, A_5, the last 'sporadic' cases corresponding to subgroups of orthogonal transformations of the 'platonic solids'. We shall not pause on this interesting fact in these notes.

(Hint for the first part: We may assume that one element of order d fixes $0, \infty$, so is of the form $t \mapsto \theta t$ for a d-th root of unity θ. Now, looking at the centralizer of this element gives the clue.)

As promised, we shall now show that every automorphism of $k(x)/k$ is of the shape illustrated above. Preliminary to this, let us recall the definition of degree of a rational function.

Definition 1.1.2. Let $r(x) \in k(x)$ be a rational function expressed as $r(x) = p(x)/q(x)$ where p, q are coprime polynomials in $k[x]$. The number $\deg r := \max(\deg p, \deg q)$ is called **degree** of $r(x)$.

Exercise 1.1.2.

(i) Observe that a rational function $r(x)$ induces a map defined everywhere from $\mathbb{P}_1(k)$ in itself. Prove that if k is algebraically closed and $\operatorname{char}(k) = 0$, for all but finitely many points $l \in \mathbb{P}_1(k)$ we have $\#r^{-1}(l) = \deg r$. (If $\operatorname{char} k = p > 0$ one has to use here the separable degree.)

(ii) Show that for a non constant $r \in k(x)$, we have $\deg r(x) = [k(x) : k(r(x))]$. (This fact shall be proved below for more general function fields in one variable. As a hint, note that x satisfies the equation $r(X) = r$, where $r = r(x)$. See also the proof of the next theorem.)

(iii) Deduce from either parts (i), (ii), that $\deg(r \circ s) = \deg r \cdot \deg s$, for any non constant rational functions, $r, s \in k(x)$. (Of course, this may be deduced also from the very definition, as given above.)

Theorem 1.1.1. *Let $r(x) \in k(x)$ be a non constant rational function. Then the extension $k(x)/k(r(x))$ is finite, of degree $\deg r$. In particular, every automorphism g of $k(x)/k$ has the shape $g(s(x)) = s(r(x))$ ($s \in k(x)$) for a unique $r = r_g \in PGL_2(k)$; equivalently, if $k(x) = k(y)$ then $y = r(x)$ for a suitable $r \in PGL_2(k)$.*

Proof. Let us write $r(x) = p(x)/q(x)$ with p, q coprime polynomials in $k[x]$, non both constant. Let us consider $F(T) := p(T) - r(x)q(T) \in k(r(x))[T]$, viewed as a polynomial in T, with coefficients in $k(r(x))$. In fact, it lies in $k[r(x), T]$. Observe that, since $r(x)$ is nonconstant (*i.e.*, not in k), $r(x)$ is transcendental over k.[4] Hence the ring $k[r(x), T]$ is isomorphic over k to the polynomial ring $k[R, T]$ in two variables. It follows that F is irreducible in $k[r(x), T]$, since it is linear in $r(x)$ and

---

[4] Indeed, we leave this implication as an easy exercise for the interested reader.

primitive. But then F is irreducible as a polynomial in T over $k(r(x))$ (here we are using Gauss' lemma).

On the other hand, as a polynomial in T, it has degree$=\max(\deg p, \deg q)$, namely it has degree $\deg r$. Since $F(x) = 0$ we obtain the first part of the theorem.[5]

As to the second part, let g be an automorphism of the field extension $k(x)/k$. Then g is determined by the value it takes on x (because $g|_k = $ id). In other words, we have $g(s(x)) = s(g(x))$. Now, $g(x)$ shall be a certain rational function $r(x)$. Therefore $g(k(x)) \subset k(r(x))$. Since $k(x)$ has degree $\deg r$ over $k(r(x))$ and since $k(r(x)) = k(x)$, since g is an automorphism, we have $\deg r = 1$. From the definition of degree, it then follows that $r(x) = l_1(x)/l_2(x)$ where l_1, l_2 are coprime polynomials of degree ≤ 1, proving completely the assertion.   □

Remark 1.1.1. It is a fact (not discussed in these notes) that automorphisms of non-rational FF are 'rare', actually finite in number execpt for the so-called *function fields of genus* ≤ 1 (the rational FF being those of genus 0).

Exercise 1.1.3. Prove that if $\sigma \in Aut_k(k(t))$ (here char$(k) \neq 2$) has order 2 then it is conjugate to the map $t \mapsto -t$.

(Hint: diagonalize a matrix representing σ, or, equivalently, consider the fixed points of σ.)

1.1.1. Unirational and rational function fields

We start with the following definition of the concept of unirational field, which corresponds to a weakened form of the rationality condition (given in Example 2.1.2).

Definition 1.1.3. We shall say that a function field in one variable K/k is **unirational** (over k) if there exists an inclusion $K \subset k(x)$ for some x (necessarily algebraic over K).

Similarly for more variables: a finitely generated extension K/k is said unirational if there exist $x_1, \ldots, x_d$ in some extension field of K,

---

[5] Alternatively, after replacing k with an algebraic closure in characteristic zero, we can argue as follows: since $F(x) = 0$, x is algebraic over $k(r(x))$ of degree $d \leq \deg r$. Let $Q(r(x), T) \in k(r(x))[T]$ be the minimal polynomial of x over $k(r(x))$, hence of degree d in T and such that $Q(r(x), x) = 0$. Let now $l \in k$ be such that $Q(l, T)$ is not identically zero and also such that the equation $r(x) = l$ has exactly $D := \deg r$ distinct roots $\rho_1, \ldots, \rho_D$ in k; note that l exists in view of Exercise 1.1.2. Then $Q(l, \rho_j) = 0$ for $j = 1, \ldots, D$, provini that $d \geq D$, and so $d = D$, as required.

algebraically independent over k, where d =trdeg(K/k), such that $K \subset k(x_1, \ldots, x_d)$.

Geometrically, if K corresponds to a variety V, this means that there is a *generically surjective* map from $\mathbb{P}_d$ to V. This viewpoint shall be briefly expanded below, in the case of function fields in one variable; see especially Remark 1.1.2(ii) and Remark 1.4.1.

Exercise 1.1.4. Prove that if $K \subset k(x_1, \ldots, x_D)$ for algebraically independent $x_1, \ldots, x_D$ and some integer $D > 0$, then one may choose D equal to the transcendence degree d of K/k, hence K/k is unirational in the above sense.

(Hint: say for instance that $D = 2$ and that K is a function field of one variable, $K = k(x, y)$, so $d = 1$. Then we can express x, y as rational functions $x(t, u), y(t, u)$ of two here independent variables. But then if we specialise *e.g.*, u to a rational function of t so that $x(t, u(t))$ remains transcendental, we are almost done. The geometrical meaning just recalled is useful for a simple proof; if D is larger than the said transcendence degree, it shall suffice to restrict the said map to a hyperplane of $\mathbb{P}_D$ transversal to a general fiber. For a complete algebraic proof, see also [31].)

Clearly every rational function field is unirational. The converse is much less obvious, and indeed has turned out not to be true in general. But it turns out to be true for function fields of one variable.[6]

This important result, due to J. Lüroth, may be proved in various ways, for instance as a consequence of the celebrated Riemann-Hurwitz formula, not proved in these notes. Here we shall offer an elementary proof, valid over any field k. (See also [22] for a proof similar to the one given here and further see [31] for an 'algorithmic' proof, based again on principles similar to the present proof. Over algebraically closed fields, a proof follows easily from the Hurwitz genus formula, once it is established that fields of genus 0 are precisely the rational fields. See further Chevalley's book [7] for a different algebraic proof.)

Theorem 1.1.2 (Lüroth). *Let k be an arbitrary field, let x be transcendental over k and let L be a field such that $k \neq L \subset k(x)$. Then there exists $r(x) \in k(x)$ with $L = k(r(x))$.*

---

[6] As to higher dimensions, the converse remains true in dimension 2 - a difficult theorem of Castelnuovo - assuming k algebraically closed. See [22] for a counterexample in dimension 2 for $k = \mathbb{R}$, due to B. Segre reproduced below in Example 1.1.4. In dimension 3 or more there are - rather deep - counterexamples even over $\mathbb{C}$, see [22] for an account of the history of this problem and for a counterexample based on the theory of quadratic forms.

Proof. We may suppose that $L \neq k$, so let $s = s(x) \in L \setminus k$, and let us write $s = a/b$ with a, b coprime polynomials in $k[x]$, not both constant. If s were algebraic over k we would have an equation $a(x) - sb(x) = 0$ for x, with coefficients algebraic over k. Since $s \notin k$, the polynomial $a(T) - sb(T)$ is not identically zero and therefore also x would be algebraic over k, against the assumptions. Hence s is transcendental over k. The same equation as above shows that x is algebraic over L, and hence $k(x)$ is a finite extension of L.[7]

Let then $F(x) = 0$ be a minimal equation for x over L, namely with $F \in L[T]$ monic, nonconstant and irreducible over L (here T is an indeterminate, *i.e.* an element transcendental over $k(x)$).

Let us write

$$F(T) = T^d + a_1 T^{d-1} + \ldots + a_d = 0, \qquad a_j \in L.$$

The coefficients $a_1, \ldots, a_d$, as elements of $L \subset k(x)$, are rational functions of x. At least one of them, say a_m, shall be nonconstant. We can then write $a_m = p(x)/q(x)$ with p, q coprime polynomials such that $\deg a_m = \max(\deg p, \deg q) > 0$. Let now $c(x)$ be a least common denominator for the coefficients $a_1, \ldots, a_d$; namely, $c(x)$ is a polynomial in $k[x]$ such that $c(x), c(x)a_1(x), \ldots, c(x)a_d(x)$ are coprime polynomials in $k[x]$. Observe that, since p, q are coprime, necessarily $q(x)$ divides $c(x)$.

From the equation $F(x) = 0$ we infer that, putting

$$G(x, T) := c(x)F(T),$$

the polynomial $G \in k[x, T]$ is primitive in x (*i.e.*, without non constant factors in x) and such that $G(x, x) = 0$.

Now, since $a_m(x) = p(x)/q(x)$, x verifies the equation $A(x) = 0$ where $A(T) := p(T) - a_m q(T) \in L[T]$. Since $F(x) = 0$ is the minimal equation of x over L we find that $F(T)$ divides $A(T)$ in $L[T]$, namely $A(T) = F(T)Q(T)$ for a $Q \in L[T]$.

Let now $c^*(x) \in k[x]$ be a minimal common denominator for the coefficients of Q (they lie in L, hence in $k(x)$). Then $R(x, T) := c^*(x)Q(T) \in k[x, T]$ is primitive in x and we have

$$c^*(x)c(x)A(T) = G(x, T)R(x, T).$$

---

[7] Of course this also follows from the fact that L has transcendence degree 1 over k and that $k(x)$ is finitely generated over k.

Observe that $q(x)A(T) = q(x)p(T) - p(x)q(T)$ lies in $k[x, T]$, so the polynomial $c^*(x)(c(x)/q(x))$ is a common factor in x to the left hand side of the last displayed equation (recall that $c(x)/q(x)$ is a polynomial). However the right hand side is primitive in x (Gauss' lemma), and hence $c^*(x)c(x)/q(x)$ is constant. In particular, $G(x, T)$ divides $q(x)p(T) - p(x)q(T)$ in $k[x, T]$. But the degree of G in x is at least $\max(\deg c(x), \deg c(x)a_m(x)) \geq \max(\deg q(x), \deg p(x))$.

Therefore we have $q(x)p(T) - p(x)q(T) = u(T)G(x, T)$, where $u \in k[T]$ does not depend on x. Now, if u were not constant in T, it would admit a root ξ in some algebraic extension k' of k and we would get $q(x)p(\xi) = p(x)q(\xi)$. Since p, q are coprime over k, they remain such also over k' and hence such equation would imply $p(\xi) = q(\xi) = 0$, contradicting again coprimality.[8]

Therefore $u(T) \in k$, proving that

$$d = \deg_T G = \deg_T(q(x)p(T) - p(x)q(T))$$
$$= \max(\deg p, \deg q) = \deg a_m.$$

Now, by Remark 1.1.1, $[k(x) : k(a_m)] = \deg a_m = d = [k(x) : L]$. Since $a_m \in L$ we have $k(a_m) \subset L$ and we finally find $[L : k(a_m)] = 1$, i.e., $L = k(a_m)$, as required. $\square$

Remark 1.1.2.

(i) The proof shows that every coefficient of $F(T)$ which is nonconstant in x is in fact a generator for L/k.

(ii) Geometrically the theorem may be interpreted as follows (see also next section). Suppose for instance that the field $K = k(u, v)$ is generated by two elements linked by an irreducible equation $f(u, v) = 0$. That K is unirational amounts to the existence of rational functions $U(x), V(x)$, not both constant, such that $f(U(x), V(x)) = 0$ identically (in fact, in such case the map $u \mapsto U(x)$, $v \mapsto V(x)$ defines an embedding of K/k in $k(x)$). For varying x in k, the curve $f(u, v) = 0$ becomes 'parametrized' by $(U(x), V(x))$ (up to finitely many exceptional values of x where U, V are not both defined).

---

[8] There are several other arguments for this simple deduction, without recourse to any extension field of k. One, that we leave to the interested readers, is to observe that $u(T)$ divides $p(T)q_i - q(T)p_i$ for p_i, q_i the coefficients of p, q, and derive a contradiction. Another one is to observe that, for x, y independent indeterminates, we have that both $q(x)p(T) - p(x)q(T)$, $q(y)p(T) - p(y)q(T)$ are divisible by $u(T)$ in $k(x, y)[T]$. But then, if $\delta = q(x)p(y) - p(x)q(y)$, we have that both $\Delta p(T)$, $\delta q(T)$ are multiples of $u(T)$, hence δ is also a multiple of $u(T)$, implying that u is constant in T.

Such parametrization shall not be bijective in general, not even up to finitely many exceptions: the points of the curve $f = 0$ shall be 'generally' recovered each a number of times equal to the degree $[k(x) : K]$; Lüroth's theorem asserts that under these assumptions we can always find a parametrization which is $1 - 1$ (again, up to finitely many exceptions).

(iii) A somewhat different proof shall be given in Example 2.6.5 below; actually, the underlying principle of that proof is similar, but there we use the theory of Valuation Rings, to derive a rather more precise conclusion.

We now reproduce an example, taken from [22] (see pages 18–20), and seemingly due to B. Segre, of a unirational but non-rational surface, defined over $\mathbb{R}$. (We shall tacitly use standard facts as the implicit function theorem.) As alluded above, over $\mathbb{C}$ no such example exists, due to a deep theorem of Castelnuovo, but (subtle and deep!) examples exist in dimension 3 or more, even over $\mathbb{C}$.

Example 1.1.4 (A unirational non-rational real surface).   We start with the cubic (elliptic) curve $Y^2 = X^3 - 3X$; since $X^3 - 3X$ has three real zeros, its real points consist of two connected components. Viewing the XY-plane as embedded in an XYZ-affine space and rotating this curve around the X-axis we obtain the surface S defined by $Y^2 + Z^2 = X^3 - 3X$. Its function field over $\mathbb{R}$ is $L : \mathbb{R}(S) = \mathbb{R}(x, y, z)$ where x, y are independent variables and $z^2 + y^2 = x^3 - 3x$.

Let us first check that S is unirational over $\mathbb{R}$, *i.e.* that L may be seen as a subfield of some field $\mathbb{R}(t, u)$. Geometrically, one may intersect S with the plane $Z = \sqrt{2}$, which produces a singular cubic; this can be parametrized by intersecting with the pencil of lines through its singular point $(-1, 0, \sqrt{2})$. Specifically one thus finds parametric formulae $x = 2 + t^2, y = t(3 + t^2), z = \sqrt{2}$ for this cubic. Now, one can rotate the cubic to get a two-dimensional parametrization for (a piece of) S; the issue is that rotations can be parametrized by points on a circle, as in Example 1.2.2 below. The final formulae are: $x = 2 + t^2$, $y = \frac{t(3+t^2)(1-u^2)-2\sqrt{2}u}{1+u^2}$, $z = \frac{2tu(3+t^2)+\sqrt{2}(1-u^2)}{1+u^2}$, which express $\mathbb{R}(S) = \mathbb{R}(x, y, z)$ as a subfield of $\mathbb{R}(t, u)$.

One can explicitly check that $[\mathbb{R}(t, u) : \mathbb{R}(S)] = 2$: see [22].

Of course by Castelnuovo's theorem mentioned above we now know that $\mathbb{C}(S)$ is rational; indeed, in this case this is very easy to establish directly: one may factor $z^2 + y^2 = (z + iy)(z - iy)$ over $\mathbb{C}$ and one gets birational formulae $x = t, 2z = u + \frac{t^3 - 3t}{u}, 2iy = u - \frac{t^3 - 3t}{u}$ (with inverses $t = x, u = z + iy$).

However, as we now prove, $\mathbb{R}(S)$ is not rational. The issue is that the set $S(\mathbb{R})$ of real points of S has two connected components (coming from the components of the original elliptic curve). Also, it may be easily checked that $S(\mathbb{R})$ consists of smooth points; through the implicit function theorem, this has the consequence that any real point of S has a two-dimensional neighborhood in S. Suppose now that $\phi : S \to \mathbb{P}_2$ is a birational map defined over $\mathbb{R}$, *i.e.* with a rational inverse $\psi : \mathbb{P}_2 \to S$, also over $\mathbb{R}$. Take points $P, Q \in S(\mathbb{R})$ lying in distinct connected components of $S(\mathbb{R})$. Since P, Q are smooth, by the above we may assume that ϕ is defined at P, Q (otherwise we move P, Q in a small neighborhood). Observe now that ψ is defined except at a finite set of points.[9] Hence we have ample choice to find a continuous connected real curve C on $\mathbb{P}_2(\mathbb{R})$, connecting $\phi(P), \phi(Q)$ and such that ψ is defined on the whole C. Then $\psi(C)$ is a real curve on $S(\mathbb{R})$ connecting P, Q, a contradiction which proves the claim.[10]

1.2. Function fields (of one variable) define (plane) curves

In this section we shall assume that k is algebraically closed.

As we have observed, it is essentially a consequence of the 'theorem of the primitive element' in algebraic field extension theory that every function field K in one variable over k is of the shape $K = k(x, y)$, where x is trascendental over k and y is then algebraic over $k(x)$. Let then $F \in k[X, Y]$ be a polynomial $\neq 0$ of minimal (positive) degree in Y such that $F(x, y) = 0$; such equation remains true on dividing by a nonzero polynomial in X and we can thus suppose that $F(X, Y)$ is primitive in X (*i.e.*, without nontrivial factors in $k[X]$). Then F shall be irreducible in $k[X, Y]$ and generates the *ideal of algebraic relations between* x, y, namely the ideal consisting of polynomials $G \in k[X, Y]$ with $G(x, y) = 0$.

Exercise 1.2.1. Prove the simple facts that we have just stated. (Hint: the irreducibility is a consequence of Gauss' lemma; for the rest, given an equation $G(x, y) = 0$, eliminate the variable Y on viewing F, G as polynomials in $k(X)[Y]$.)

---

[9] This holds for every rational map on a smooth surface, and follows from the fact that the local ring at smooth points is a UFD. However in the present case of $\mathbb{P}_2$ a direct easy argument works: a point at which ψ is not defined is a zero of certain two fixed coprime polynomials in two variables.

[10] Work of Colliot-Thélène and others has proved that in particular that $k(S)$ is not rational if k is a number field not containing $\sqrt{-1}$.

Now, we may consider as associated to F the *affine plane curve* $C_F :=$ $\{(a, b) \in k^2 : F(a, b) = 0\}$.

In this way, to every description of K by means of two generators we may associate a plane curve. As in the Introduction, we speak of such a curve as a *model* of K. If C is such a curve, we write $K = k(C)$.

(Note also that all polynomials cF^m, $c \in k^*, m > 0$, define the same curve.)

Exercise 1.2.2. Prove that such curve is not a nontrivial union of curves of the same type, namely it is *irreducible*. This follows from the irreduciblity of F. (Recall that k is supposed here to be algebraically closed.)

Considering the homogeneized polynomial $\tilde{F}(X, Y, Z) :=$ $Z^{\deg F} F(X/Z, Y/Z)$, we obtain the projective plane curve $\tilde{C}_{\tilde{F}} := \{(a : b : c) \in \mathbb{P}_2(k) : \tilde{F}(a, b, c) = 0\}$. Those of its points which have $c \neq 0$ correspond bijectively with the points in C_F. On the other hand, the points with $c = 0$ form a finite set with at most $\deg F$ elements; they are called *points at infinity* (relative to the immersion of the affine plane $\mathbb{A}^2$ in $\mathbb{P}_2$ which identifies a point $(a, b) \in \mathbb{A}^2(k) = k^2$ with $(a : b : 1) \in \mathbb{P}_2$).

Conversely, let $F \in k[X, Y]$ be an irreducible polynomial defining as above the plane affine curve C_F. Then F generates a prime ideal (F) in $k[X, Y]$, so the quotient ring $k[X, Y]/(F)$ is a domain, denoted $k[C_F]$. The fraction field K of such domain is then a function field in one variable over k, in the previous meaning. (We have $K = k(C_F)$.)

Exercise 1.2.3. Prove the last statement.

Even starting with a description of K/k given by any number n of generators, we would obtain corresponding geometric models, embedded in a space of dimension n: if $K = k(x_1, \ldots, x_n)$, consider, to start with, the ideal $I \subset k[X_1, \ldots, X_n]$ formed by the polynomials F for which $F(x_1, \ldots, x_n) = 0$, indeed called the *ideal of relations*. Since K is a domain, such ideal is prime and defines an irreducible curve $C \subset k^n$ as $C := \{(a_1, \ldots, a_n) \in k^n : F(a_1, \ldots, a_n) = 0 \; \forall F \in I\}$.[11] Even now, we can consider the *projective completion* $\tilde{C}$ of the curve; this is defined as the set of points $(a_0 : a_1 : \ldots : a_n) \in \mathbb{P}_n(k)$ such that $\tilde{F}(a_0, \ldots, a_n) = 0$ for all homogeneous polynomials $\tilde{F}(X_0, \ldots, X_n)$ such that $\tilde{F}(1, X_1, \ldots, X_n) \in I$.

---

[11] Saying that C is a 'curve' is, by definition, like saying that K/k has transcendence degree 1.

Remark 1.2.1.

(i) It is important to note that for a fixed field K we obtain in this way different curves (*i.e.*, different models of K), depending on the system of generators that we use to describe the extension K/k. We shall see several examples of this. Yet, the fact that these curves are models of the same field K, *i.e.*, have the same function field K, reflects in their *birational equivalence*. This notion will be made more precise in the sequel.

(ii) It is also important to note that if we consider the 'ideal of the curve', *i.e.* $J_C := \{f \in k[\mathbf{X}] : f(P) = 0 \; \forall P \in C\}$, we obtain back the ideal I. Indeed, that $J_C \supset I$ is clear, but in fact, due to the celebrated 'Nullstellensatz' of Hilbert (see Appendix A below for a discussion) we have $J_C = \sqrt{I}$; however I is prime and hence $\sqrt{I} = I$, as wanted.

(iii) Since the above defined ideal I is prime, we can consider the fraction field Γ of the domain $A := k[X_1, \ldots, X_n]/I$. In fact, we have that Γ is isomorphic to K over k. To prove this, define a homomorphism over k from A to $k[x_1, \ldots, x_n]$, by sending X_i modulo I to x_i. By definition of I, this is well defined and an isomorphism, which implies at once the assertion.

Example 1.2.1 (Models of rational fields).   When K is rational, an equality $K = k(x)$ corresponds to a model of K given by a line. In fact, we have a single generator, and there are no algebraic relations involving it. This line naturally may be also immersed in higher dimensions. For instance, the description of K with two (equal) generators $K = k(x, y)$, where $y = x$, corresponds to the model of a line given by $Y = X$ in the affine plane.

Example 1.2.2 (More rational fields, conics and singular cubics).
Interesting examples of rational plane curves given by equations of degree > 1 are as follows.

A typical one comes from the 'unit circle' with equation $X^2 + Y^2 = 1$; this defines a function field $k(x, y)$, where $y^2 = 1 - x^2$. Projection from a point on the circle onto a line not passing through the point yields formulae showing that K is rational. (See Exercise 1.4.1 below; the method applies to every irreducible plane curve defined by a quadratic polynomial, *i.e.* to every *conic*.)

Next, consider the polynomial $Y^2 - X^3 - cX^2$, $c \in k$, which is again irreducible. It defines a function field $K = k(x, y)$, where x is transcendental over k and $y^2 = x^3 + cx^2$. Putting $t = y/x \in K$, we have $t^2 = x + c$, hence $x = t^2 - c$ and $y = tx = t^3 - ct$. Therefore $K = k(t)$ is rational. (Note that the origin is a *singular point* of the curve defined by the equation, *i.e.* the tangent space coincides with the ambient space and hence has dimension greater than 1. Geometrically, the substitution corresponds to intersecting the cubic with a line through the origin; since the line meets the cubic in three points, and two intersections are absorbed by the singular origin, we have just another intersection, providing a $1 - 1$ correspondence of the set of lines - *i.e.* $\mathbb{P}_1$ - with the cubic.)

Example 1.2.3 (Elliptic fields). The 'cubic' function field of the previous example comes from the equation $y^2 = f(x)$, where f is a cubic polynomial, with the multiple root $x = 0$ (double or triple according as $c \neq 0$ or not). Suppose now that f is still cubic, but with no multiple factors. Then the function field $K = k(x, y)$, where $y^2 = f(x)$ is called usually **elliptic**, because of the deep relationship with elliptic function theory (which unfortunately we shall not have the time to touch at all here). However, let us pause to prove that they are never rational fields, and not even unirational (and without using Lüroth's theorem for this last assertion).

To start with, let us observe that the substitution $u = 1/x, v = y/x^2$ is invertible: indeed, we have $x = 1/u, y = vx^2 = v/u^2$. Therefore $K = k(x, y) = k(u, v)$ (we have here an example of birational correspondence, see also below).

The elements u, v now verify the equation $v^2 = u^4 f(1/u) = g(u)$, where $g(T) := T^4 f(1/T)$ is a polynomial of degree 3 or 4 without multiple roots; the degree is 3 precisely when $f(0) = 0$, but we can suppose that this is not the case by performing a translation on x. It shall suffice therefore to work with fields defined by an equation of this type and degree 4.

Then, the above assertion concerning non-(uni)rationality is the content of the following theorem, which we display.

Theorem 1.2.1. *Let $v^2 = g(u)$ where u is transcendental over k and where $g \in k[T]$ is a polynomial of degree 4 without multiple factors. Then the field $K = k(u, v)$ is not unirational over k.*

For the proof, let us assume by contradiction that $k(u, v)$ is contained in $k(z)$ for some element z, algebraic over $k(u, v)$. Then we could write u, v as non constant rational functions $u(z), v(z) \in k(z)$ such that $v^2(z) = g(u(z))$. Thus the theorem shall be a consequence of the following result, formulated in terms of a diophantine equation over a function field:

Theorem 1.2.2. *Let $g \in k[T]$ be a polynomial of degree 4 without multiple roots. Then the equation $v^2 = g(u)$ has no solutions in nonconstant rational functions $u, v \in k(z)$.*

Proof. Let by contradiction $(u(z), v(z)) \in k(z)^2$ be a non constant solution. If we write $u(z) = a(z)/b(z)$, for coprime $a, b \in k[z]$, then the denominator of $g(u(z))$ shall be $b^4(z)$, so that $v(z) = c(z)/b^2(z)$, for a suitable polynomial $c \in k[z]$.

We shall employ 'Fermat's descent'. We can assume that our example is such that $\deg u(z) = \max(\deg a, \deg b)$ is positive and minimal among all such situations.

Write now $g(T)$ in the shape $g_0 \prod_{i=1}^4 (T - l_i)$, where $g_0 \in k^*$ and where $l_1, \ldots, l_4 \in k$ are pairwise distinct. Then $c^2(z) = g_0 \prod_{i=1}^4 (a(z) - l_i b(z))$. Note that, since $a(z), b(z)$ are coprime and the l_i are distinct, the factors on the right hand side are pairwise coprime (a common root to two of them would a root of both a and b).

At this stage, the fundamental remark is that *since the product of the coprime factors in question is a square in $k[z]$, then each factor must be in fact a square*. Indeed, if a factor had a root of odd multiplicity, this would be a root also of another factor (which is impossible), since it appears in the product with even multiplicity. Hence each factor has only roots of even multiplicity and is therefore the square of a polynomial, because the constant field k is algebraically closed.

Then we can write

$$a(z) - l_i b(z) = q_i^2(z), \qquad q_i \in k[z], \qquad i = 1, 2, 3, 4,$$

where the q_i are suitable polynomials which must be pairwise coprime (for $a(z), b(z)$ are coprime) and nonzero. Eliminating a, b from the first two equations and substituting in the remaining two ones, we easily obtain

$$(l_i - l_1)q_2^2(z) - (l_i - l_2)q_1^2(z) = (l_2 - l_1)q_i^2(z), \qquad i = 3, 4.$$

Let now $h(T) := ((l_4 - l_1)T^2 - (l_4 - l_2))((l_3 - l_1)T^2 - (l_3 - l_2))$; it is verified at once that h has degree 4 and has no multiple factors. Putting then $r(z) := q_2(z)/q_1(z), s(z) := (l_2 - l_1)q_3(z)q_4(z)/q_1^2(z)$, we obtain, on multiplying term by term the displayed equations for $i = 3, 4$, that $s^2(z) = h(r(z))$.

We have then reached a situation analogous to the previous one, with h in place of g and with r, s in place of u, v. Note indeed that r cannot be constant, because q_1, q_2 are coprime and not both constant.

However, now the degree in question is decreased. Indeed, we have

$$\deg r \leq \max(\deg q_1, \deg q_2)$$
$$\leq \max(\deg a, \deg b)/2 = (\deg u)/2 < \deg u.$$

But this is a contradiction with the minimality of $\deg u$ among all possible counterexamples, which finally proves the assertion.   □

Remark 1.2.2.

(i) In this proof an analogy clearly appears between *diophantine* problems in classical sense[12] and certain problems of geometric nature, such as for instance the one considered above (of proving the non(uni)rationality of a specific curve). In fact, some geometric problems may be considered as diophantine problems to be solved inside a function field in place of $\mathbb{Q}$: for instance here we had to prove that *the equation* $v^2 = g(u)$ *has no solution in rational functions* $u, v \in k(z)$.

(ii) It is an interesting exercise to study the various function fields which appear in the given proof, and their relationships. Let us define the field $L = k(t_1, t_2, t_3, t_4)$ by putting $t_i^2 = u - l_i, i = 1, 2, 3, 4$. Hence $L \supset k(u)$ (u here is transcendental over k). Define now $r_i = t_i/t_1$ and put $F := k(r_2, r_3, r_4)$. Clearly, we have $F \subset L$ and we see that $[L : F] = 2$. The r_i verify $(l_i - l_1)r_2^2 - (l_i - l_2) = (l_2 - l_1)r_i^2$. Let us also observe that $r_2^2 = (u - l_2)/(u - l_1)$, whence $k(u) = k(r_2^2)$. Further, let $s := (l_2 - l_1)r_3r_4$. Then we see at once that $s^2 = h(r_2)$, where h is as in the proof of Theorem 1.2.1; put $H = k(r_2, s)$, so that $H \subset F$. Finally, putting $v = \sqrt{g_0}(u - l_1)^2r_2r_3r_4$ one verifies that $v^2 = g(u)$ and it is clear that $K = k(u, v) \subset H$. We have $[k(r_2) : k(u)] = 2$, whence $[H : K] = 2$.

The above proof works since an inclusion $K \subset k(z)$ induces (and here we use unique factorization in $k[z]$) an inclusion $F \subset k(z)$ (in fact, the r_i 'become' the $q_i(z)/q_1(z)$). Hence also $H \subset k(z)$, which yields the 'descent', since the degree is decreased: $[k(z) : K] = [k(z) : H][H : K]$. (Going ahead with the descent, *i.e.* repeating the construction of the chain of fields, with H in place of K, one would find that the new intermediate field H', with $H \subset H' \subset k(z)$, is isomorphic with K (over k); this may be explained with the duality for the so-called isogenies of elliptic fields. (See [34].)

Topological viewpoint: It may be not free of interest to review the proof in topological terms, relying on a few necessary basic concepts.

With this in mind, note that the field inclusion $K \subset \mathbb{C}(z)$ corresponds to a rational nonconstant map $\varphi : \mathbb{P}_1 \to E_K$, where $\mathbb{P}_1$ denotes here the set $\mathbb{P}_1(\mathbb{C})$ of complex points, *i.e.* the Riemann sphere S_2 and similarly

---

[12] In such problems it was originally required to describe the solutions in integer or rational numbers, to (usually algebraic) equations. Naturally this archaic terminology has evolved, to include other problems and concepts.

$E_K = E_K(\mathbb{C})$ is (the projective completion of) the complex curve $v^2 = g(u)$. Similarly, the inclusion $K \subset H$ corresponds to a map $\psi : E_H \to E_K$, where E_H is the curve defined by $s^2 = h(r_2)$. Now, we see that ψ defines a cover (there is no ramification) of degree 2, so that ($\mathbb{P}_1$ is simply connected) the map φ lifts to a map $\sigma : \mathbb{P}_1 \to E_H$ with $\psi \circ \sigma = \varphi$. It follows that $\deg \varphi = 2 \deg \sigma$ and we thus have the sought 'descent' (since absence of ramification also shows that σ is itself a rational map.)

(iii) It is a bit simpler to prove that an elliptic field $K = k(x, y)$ is not rational (rather than unirational - here of course we think of arguments which avoid Lüroth Theorem 1.1.2). Here is a possible argument. If K were rational, there would exist nonconstant rational functions $x(t), y(t) \in k(t)$ such that $y(t)^2 = f(x(t))$ (here we are using only unirationality) and moreover such that $k(t) = k(x(t), y(t))$. Now, K is isomorphic over k to $k(x(t), y(t))$, through $x \mapsto x(t)$, $y \mapsto y(t)$, and if K is rational we may suppose that equality holds between these fields. Then, the automorphism of K which fixes $k(x)$ and changes the sign of y would correspond to an automorphism of $k(t)$ of order 2. By Exercise 1.1.3, on changing t with a suitable $\sigma(t), \sigma \in PGL_2(k)$, we can then assume that such automorphism is given by $t \mapsto -t$. Moreover, since $t \in K$, we can write $t = a(x) + yb(x)$, with $a, b \in k(x)$. But $t^2 \in k(x)$ (because $k(x)$ is the fixed field of the automorphism in question) and we deduce at once that $a(x) = 0$. Then $y = t/b(x)$ and substituting into the fundamental equation, we have $t^2 = b(x(t))^2 f(x(t))$. Finally, since $x(t)$ is invariant by $t \mapsto -t$, we shall have $x(t) = r(t^2)$. Substituting into the last equation and setting $u := t^2$ we obtain $u = b(r(u))^2 f(r(u))$, whence $u \in k(r(u))$. Then $s := r(u)$ has degree 1 (as a rational function of u) and u is of the shape $l(s)$, with $\deg l = 1$, whence $l(s) = b(s)^2 f(s)$ identically in s, and now we may easily obtain a contradiction with the fact that f has no multiple roots.

Exercise 1.2.4.

(i) Let f, g be nonconstant coprime squarefree polynomials in $k(x)$. Let us define $K = k(x, y, z)$, where x is transcendental over k and where $y = \sqrt{f(x)}, z = \sqrt{g(x)})$ (for some determination of the square roots).

 (a) Prove that $[K : k(x)] = 4$.
 (b) Prove that the ideal of relations among x, y, z is generated by $Y^2 - f(X)$ e $Z^2 - g(X)$.

(Hint: Let J be the ideal generated by the polynomials in question and assume that $F(x, y, z) = 0$. Then $F(x, y, Z) = Q(x, y, Z)(Z^2 - g(x))$ for a $Q \in k(x, y)[Z]$; since $Z^2 - g(x)$ is monic, in fact $Q \in k[x, y][Z]$. Repeating the argument for $F(x, Y, Z) - Q(x, Y, Z)(Z^2 - g(x))$, with Y in place of Z one concludes. See next exercise for generalizations.)

Deduce that the space curve $\{(a, b, c) \in k^3 : b^2 = f(a), c^2 = g(a)\}$ is irreducible and yields a model of K.

(c) Prove that $K = k(x, y + z)$ and construct a corresponding plane model for K. (It will suffice to seek an irreducible equation for $y + z$ over $k(x)$.)

(ii) Let $f, g \in k[X, Y]$ be irreducible, not constant in Y and such that if $f(x, y) = 0$ and $g(x, z) = 0$ (where $x \notin k$) then $k(x, y)$ and $k(x, z)$ are linearly disjoint over $k(x)$. (In pratice we are asking that $g(x, Z)$ is irreducible in $k(x, y)[Z]$.)

(a) Prove that the field $k(x, y, z)$ is well-defined up to isomorphism over k and is a function field of one variable over k.

(b) Prove that the ideal of relations among x, y, z coincides with the ideal J generated by $f(X, Y), g(X, Z)$.

(Hint: similarly to the previos exercise, prove that if $F(x, y, z) = 0$ then $c(X)F(X, Y, Z) \in J$ for a suitable nonzero $c \in k[X]$. Reduce then to prove the implication $(X - l)F \in J \implies F \in J$ for $l \in k$. If for instance $l = 0$, let $XF = Af + Bg$; summing to F an element in J one can assume that A, B do not depend on X. Now, for $X = 0$, $A(Y, Z)f(0, Y) + B(Y, Z)g(0, Z) = 0$, whence $A = Qg(0, Z)$, $B = -Qf(0, Y)$ for a suitable polynomial Q. Finally, from $A = Qg(X, Z) + XA_1, B = -Qf(X, Y) + XB_1$ one may conclude.)

1.3. Algebraic varieties, rings of regular functions

In this section, in the same line as above, we only recall very basic concepts , mainly definitions, about algebraic varieties of any dimension. For more comprehensive introductions, see e.g. [9, 14, 30].

As above, k shall be usually supposed to be algebraically closed.

Let J be an ideal in $k[\mathbf{X}] := k[X_1, \ldots, X_n]$. Then J determines an *affine variety* $V = V_J := \{P \in k^n : F(P) = 0 \,\forall F \in J\}$[13] It is a simple but fundamental fact that such varieties constitute the set of *closed sets* for a topology, denominated the *Zariski topology*.

---

[13] For simplicity we shall always identify a variety with the set of its k-points.

Coversely, given such a set $V = V_J$, we can consider the ideal $J^* :=$ $\{F \in k[\mathbf{X}] : F(P) = 0 \ \forall P \in V\}$. In general, we have $J \subset J^*$ and the celebrated *Nullstellensatz* of Hilbert (see Appendix A for a proof) states that $J^* = \sqrt{J} := \{F : \exists m \in \mathbb{N}, F^m \in J\}$.

It is not difficult to prove (**Exercise**) that if J is prime then V is irreducible (namely, it is not the union of two smaller varieties of the same shape); from the Nullstellensatz it also follows the equivalence between this irreducibility and the primality of $\sqrt{J}$.

Recall also (Hilbert basis theorem) that every finitely generated algebra over k (*i.e.* of the shape $k[x_1, \ldots, x_n]$) is a Noetherian ring (*i.e.*, every ideal is finitely generated). It follows that V may be always expressed (uniquely) as a finite union of maximal irreducible subvarieties, called *components*.

Suppose now that J is prime (so, in particular, $J = \sqrt{J}$). Then the quotient $k[V] := k[\mathbf{X}]/J$ is an integral domain, whose fraction field $K = k(V)$ is denominated *field of rational functions on V* (over k). Let x_i be the image of X_i in $k[V]$. An element φ in K may be written (in several ways) as a quotient $R(x_1, \ldots, x_n)/S(x_1, \ldots, x_n)$, where $R, S \in k[X_1, \ldots, X_n]$.

Observe that an element $f \in k[V]$ defines a function on V as follows: let $F \in k[\mathbf{X}]$ be a polynomial whose class modulo J is f. Then we set $f(P) := F(P)$ for a $P \in V$, this value being independent of the representative F. On the other hand, the elements of $k(V)$ define only functions restricted to their *regularity set*, or *domain of definition*, defined later.

The number $\operatorname{tr} \deg(K/k)$ is called *dimension of V*; when V has dimension 1, V is said an (affine) *algebraic curve* and K is a function field in *one variable* in the same sense considered above.

In the case when the ambient space is projective space $\mathbb{P}_n(k)$, rather than affine space $\mathbb{A}_n(k) = k^n$, one can develop analogous notions. An important difference is that a vanishing condition $F(a_0, \ldots, a_n) = 0$, for a polynomial $F \in k[X_0, \ldots, X_n]$, is well defined for the point $P = (a_0 : a_1 : \ldots : a_n) \in \mathbb{P}_n(k)$ only if it does not depend on the projective coordinates of P; now, one can easily see that this happens only if all the homogeneous components of F vanish at P. Hence in the projective case the algebraic sets shall be defined by *homogeneous ideals*, namely those ideals which can be generated by homogeneous polynomials. For an algebraic set V, we can still define a 'coordinate ring' $k[V] := k[X_0, \ldots, X_n]/J_V$, but now the elements of $k[V]$ shall not define functions on V (because the value of a polynomial depends on the choice of projective coordinates). However the ratio of two elements of $k[V]$ represented by homogeneous polynomials of the same degree has a well-defined value on the set of points where the denominator does not vanish. The set of such ratios constitutes the *field of rational functions* $k(V)$.

Often, one may reduce issues in projective space to the affine case on covering projective space with *open affine* sets, for instance those defined by the nonvanishing of a given coordinate.

Going back to the affine case, if $P \in V$ we define the *ring of regular functions at P* as

$$\mathcal{O}_P := \{\varphi \in K : \exists R, S \in k[V], \ \varphi = R/S, \ S(P) \neq 0\}. \qquad (1.1)$$

Note that the condition $S(P) \neq 0$ may possibly hold only for a subset of the representations of φ as a ratio R/S. For such representations the value $\varphi(P) = R(P)/S(P)$ is well defined.

Observe also that $\mathcal{O}_P$ is a local ring whose maximal ideal is made up of the functions φ with $\varphi(P) \neq 0$. A given function $\varphi \in K$ in general may be evaluated at the points P such that $\varphi \in \mathcal{O}_P$. It is not difficult to prove that this holds for all points except those $P \in W = W_\varphi$, where $W \subset V$ is a variety of dimension smaller than $\dim V$. In the case when V is a curve, W is therefore a finite set.

Exercise 1.3.1.

(i) Prove directly the last statement for a plane curve.
  (Hint: Observe that if $F(X,Y)$ is an irreducible polynomial which does not divide $G(X,Y)$, then the system $F(a,b)=G(a,b)=0$ has only finitely many solutions in k^2; for this, use for instance an equation $A(X,Y)F(X,Y)+ B(X,Y)G(X,Y) = 1$, $A, B \in k(X)[Y]$.)
(ii) Prove that a plane curve contains infinitely many points.

1.4. Field inclusions and rational maps

Let $K \subset L$ be function fields in one variable over k. Since $K/k, L/k$ both have transcendence degree 1, L is algebraic over K. Moreover L is finitely generated over k, so *a fortiori* it is such over K, whence L/K is an extension of finite degree $[L : K]$.

Suppose we are given systems of generators for K, L, namely equalities $K = k(x_1, \ldots, x_m), L = k(y_1, \ldots, y_n)$. Then the inclusion $K \subset L$ amounts to the existence of rational functions $\varphi_1, \ldots, \varphi_m \in k(T_1, \ldots, T_n)$ such that $x_i = \varphi_i(y_1, \ldots, y_n)$.[14]

---

[14] Here is is understood that the denominator of the φ_i does not vanish at $(y_1, \ldots, y_n)$. Moreover, it is clear that the φ_i are not unique: if $\psi(y_1, \ldots, y_n) = 0$, then φ_i may be replaced by $\varphi_i + \psi$.

Now, if $F(x_1, \ldots, x_m) = 0$ is a relation among the x_i, by substitution we obtain the equation $F(\varphi_1(y_1, \ldots, y_n), \ldots, \varphi_m(y_1, \ldots, y_n)) = 0$. The numerator of the rational function $F(\varphi_1, \ldots, \varphi_m)$ thus corresponds to a relation among the y_j. (In other words, composition with the φ_i sends the relations among the x_i to relations among the y_i.)

Let now C_K, C_L models for K, L, namely curves respectively in k^m, k^n, associated to K, L as in Section 1.2.[15] Then, taking into account what we have just observed on the relations among the x_i and the y_i, it is not difficult to verify that the inclusion $K \subset L$ corresponds to a *nonconstant rational map* $\varphi = (\varphi_1, \ldots, \varphi_m)$ from C_L to C_K; namely, this map sends a point $(a_1, \ldots, a_n) \in C_L$ in $(\varphi_1(a_1, \ldots, a_n), \ldots, \varphi_m(a_1, \ldots, a_n)) \in C_K$. It is important to note that such φ is not quite a function defined on the whole curve C_L: it shall be generally defined only on the points $P \in C_L$ such that $\varphi_1, \ldots, \varphi_m \in \mathcal{O}_P$. For these points we shall have, however, $(\varphi_1(P), \ldots, \varphi_m(P)) \in C_K$. Such set of points is anyway 'very' vast, in the sense that it consists of the whole C_L with only finitely many exceptions. (See the previous section.)

Suppose for simplicity that $C_K = \mathbb{P}_1$ so that $K = k(x)$ for an x trascendental over k.

Suppose also that L/K is separable. Then, by the theorem of primitive element, we may write $L = k(x, y)$, where y verifies an irreducible algebraic equation

$$P(x, y) = y^d + a_1(x)y^{d-1} + \ldots + a_d(x) = 0, \quad a_i \in K.$$

The rational map φ now corresponds to the projection $\varphi(x, y) = x$, and we have $d = [L : K]$.

Note that $P(x, y)$ is irreducible in y and by separability has not multiple factors. But then, for all but finitely many $x_0 \in k$, $P(x_0, y)$ has not multiple factors and thus shall have d distinct roots $y_1, \ldots, y_d$. Then the fiber $\varphi^{-1}(x_0)$ consists of $\{(x_0, y_1), \ldots, (x_0, y_d)\}$ and hence $\#\varphi^{-1}(x_0) = d = [L : K]$.

We can resume these facts in the following theorem-definition:

Theorem 1.4.1. *Let $K = k(x_1, \ldots, x_m) \subset L = k(y_1, \ldots, y_n)$ be two function fields in one variable over k, and let C_K, C_L be two plane curves, models of K, L respectively, corresponding to certain given systems of generators. Then the map $\varphi : C_L \to C_K$ constructed above is defined at all but finitely many points of C_L. Moreover, if the extension L/K is*

---

[15] Of course they depend on the chosen systems of generators, not only on the fields K, L, but for simplicity we do not make it explicit this dependence.

separable for all points $P \in C_K$ *with finitely many exceptions, we have* $\#\varphi^{-1}(P) = [L : K]$. *This number is called* **degree** *of the map* φ.

Example 1.4.1. When $K = k(x)$ is a rational field, generated by the single element x, we have that C_K is a line, *i.e.* $\mathbb{P}_1$. We have $K \subset L$ if and only if $x \in L$; hence any element x in $L \setminus k$ individuates, as above, a rational map from C_L to $\mathbb{P}_1$.

When we have an equality $K = L$, but different systems of generators, then the curves C_K e C_L (which depend, as remarked above, on the generators, not only on the fields) are different models of the same function field K. The rational map φ shall have an inverse $\psi : C_K \to C_L$, also rational and defined outside a finite subset of C_K. We then say that C_L e C_K are *birationally equivalent*.

Example 1.4.2. Consider the field of Example 1.2.3, defined by an equation $y^2 = f(x)$, where f is a cubic polynomial without multiple roots. The curve $C = \{(a, b) \in k^2 : b^2 = f(a)\}$ is a model of the field $K = k(x, y)$, where $y^2 = f(x)$. Put (as in the said example) $u = 1/x$, $v = y/x^2$. Then the field $L = k(u, v)$ equals K: in fact, we have directly $L \subset K$ and moreover $x = 1/u$, $y = vx^2 = v/u^2$, proving the opposite inclusion. The elements u, v now verify the equation $v^2 = u^4 f(1/u) = g(u)$, where $g(T) := T^4 f(1/T)$ is a polynomial of degree 3 (if $f(0) = 0$) or 4. The curve $C' = \{(a, b) : b^2 = g(a)\}$ is a model of L, and hence also of K. The corresponding rational map ψ from C to C' is given by the above substitution: if $(a, b) \in C$, then $\psi(a, b) = (1/a, b/a^2)$. This map is defined for every point such that $a \neq 0$. its inverse is also expressed by $(a, b) \mapsto (1/a, b/a^2)$.

Remark 1.4.1. If L is unirational, then $L \subset K := k(z)$ for a z algebraic over L; now there exists a rational nonconstant map from a line (which is a model of K) to any model C of L. One then speaks of a *parametrization* of C (see also the above remarks to Lüroth's theorem); for instance, if C is defined by $f(x, y) = 0$, there exist rational functions $x(z), y(z)$, not both constant, such that $f(x(z), y(z)) = 0$ identically. In such a situation the points of C can be obtained (up to finitely many exceptions) simply by substituting for z elements of k. Finally, note that Lüroth's theorem may be rephrased by saying that *if there exists a parametrization, then there exists one which is bijective, up to finitely many points*. Indeed, a parametrization implies that L is contained in a rational field, so it is itself rational, equal to $k(x)$ for some x; then the parametrization corresponding to the inclusion $L \subset k(x)$ is invertible, proving the said bijectivity.

Exercise 1.4.1.

(i) Notation being as in Exercise 1.2.4, consider the models of K given by $K = k(x, y, z)$ and $K = k(x, y + z)$, corresponding to curves C_1, C_2 in three and two dimensions respectively. Construct explicitly inverse rational maps between these curves and provide finite sets of points such

(ii) Let $d \geq 2$ be an integer and let $K = k(x, y)$ where $x^d + y^d = 1$. The corresponding model of K is usually called *Fermat curve* of degree d.

    (a) Show that for d coprime to char(k), $X^d + Y^d - 1$ is irreducible over k, hence K is indeed well defined up to isomorphism over k. (There are several ways to proceed; for instance, one can consider the roots of the polynomial with respect to the variable Y. Or one can use the invariance with respect to the substitutions $X \mapsto \theta X, Y \mapsto \eta Y$, for θ, η and d-th roots of unity, or also Eisenstein criterion applied to the ring $k[X]$, or even further arguments...)

    (b) For $d = 2$, write explicitly an element $t \in K$ with $K = k(t)$, obtaining a bijective correspondence with a line. (Now the curve is a circle and one can project it onto a line, from a point on it. For instance, one can choose the Y-axis as the line and $(-1, 0)$ as the point. Then, on letting t be the slope of the variable line through $(-1, 0)$ giving the projection, one obtains the well-known formulas $x = (t^2 - 1)/(t^2 + 1)$, $y = 2t/(t^2 + 1)$ and the inverse $t = y/(x + 1)$.)

    (c) For $d \geq 3$ (and char$k = 0$) prove directly that K is not unirational.
(Hint: otherwise there would exist nonconstant rational functions $f(t), g(t)$ with $f^d + g^d = 1$. Differentiating one gets $f'(t) f^{d-1}(t) + g'(t) g^{d-1}(t) = 0$. But f, g cannot have common zeros; hence conclude on counting degrees and zeros. Another proof of non-rationality comes from Exercise 1.1.1(vi), on noting that the present field K has automorphisms $x \mapsto \theta x, y \mapsto \eta y$ forming a group isomorphic to the direct product of two cyclic groups of order d.)

Let us consider now two irreducible polynomials $F_1, F_2 \in k[X, Y]$, and let us associate to them function fields $K_1 = k(x_1, y_1)$, $K_2 = k(x_2, y_2)$, where as usual x_1, x_2 are transcendental over k and where $F_i(x_i, y_i) = 0$, $i = 1, 2$. The polynomials F_i define plane curves C_i.

It may happen that for instance K_1 is isomorphic (over k) to a subfield of K_2. We can then identify K_1 with such a subfield of K_2, so that we shall be able to write x_1, y_1 as rational functions $x_1 = \xi(x_2, y_2)$, $y_1 = \eta(x_2, y_2)$, where $\xi, \eta \in k(x_2, y_2) = K_2$.

Let us pause at once to observe that ξ, η may be thought of as represented by rational functions in $k(X, Y)$, evaluated at (x_2, y_2). For this, it shall be necessary that their denominator is not divisible by $F_2(X, Y)$; moreover, of course such representative functions shall not be unique (for we may add to them arbitrary multiples of F_2).

Naturally, we must have $F_1(\xi(x_2, y_2), \eta(x_2, y_2)) = 0$, which in turn amounts to the fact that the numerator of $F_1(\xi(X, Y), \eta(X, Y))$ is divisible by $F_2(X, Y)$. It is then easy to see that if $(a, b) \in k^2$ is a point of C_2, and if ξ, η are defined at (a, b), then $(\xi(a, b), \eta(a, b))$ is a point of C_1. In other words, the inclusion of fields determines a rational map (ξ, η) from C_2 to C_1.

If moreover the said isomorphism is surjective, and then identifies K_1 and K_2, there eixists correspondingly also an inverse rational map. Then, as above, we may speak of *birationally equivalent curves* C_1, C_2 (over k). They correspond to polynomials F_1, F_2 which give different (plane) models of a same function field $K = K_1 = K_2$.

As mentioned above, these rational maps are not defined everywhere in general (like for the usual rational functions). One can see that this problem, however, is relatively mild, in the sense that it happens only for a finite number of points. Moreover, passing to projective smooth models, the rational maps become defined everywhere, a phenomenon that does not happen on higher dimensions, and renders the theory of curves much easier than the case of general varieties.

We conclude here our very brief discussion of algebraic sets, mainly of dimension one. In the next chapter we shall study valuation rings; such theory may be applied to a more careful study of function fields in one variable, but we postpone such applications to an expanded version of the present notes.

Chapter 2
Valuation rings

2.1. Valuation rings and places

Definition 2.1.1. Let K be a field. We say that a subring $\mathcal{O} \subset K$ is a *valuation ring (abbr. VR) of K* if

(i) $\mathcal{O} \neq K$, and

(ii) for every $x \in K \setminus \mathcal{O}$ we have $x^{-1} \in \mathcal{O}$.

In case k is a subfield of K we say that '$\mathcal{O}$ is a VR of K/k' if moreover $k \subset \mathcal{O}$.

The set of valuation rings of K (resp. of K/k) will be denoted by Σ_K (resp. $\Sigma_{K/k}$).

Note that a VR $\mathcal{O}$ of K is in particular a domain with fraction field K.

Proposition 2.1.1. *For $\mathcal{O} \in \Sigma_K$ the set $\mathcal{P} := \mathcal{O} \setminus \mathcal{O}^* = \{x \in \mathcal{O}, x^{-1} \notin \mathcal{O}\}$ is a maximal nonzero ideal of $\mathcal{O}$. In particular, $\mathcal{O}$ is a local ring and $\mathcal{P}$ is its unique maximal ideal. We shall also refer to the pair $(\mathcal{O}, \mathcal{P})$ as a VR of K.*

The quotient ring $\mathcal{O}/\mathcal{P}$ is therefore a field, called the residue field. If $\mathcal{O} \in \Sigma_{K/k}$, the field k embeds isomorphically in $\mathcal{O}/\mathcal{P}$ under the quotient map.

Proof. To start with, if $\mathcal{P}$ were $\{0\}$, $\mathcal{O}$ would be a field, necessarily $= K$, a contradiction. Let $x \in \mathcal{P}$ and $y \in \mathcal{O}$. Then $xy \notin \mathcal{O}^*$ for otherwise for some $z \in \mathcal{O}$ we would have $xyz = 1$ and x would also have an inverse in $\mathcal{O}$. This proves that $xy \in \mathcal{P}$.

Let then $x, y \in \mathcal{P} \setminus \{0\}$. Then either $y/x \in \mathcal{O}$ or $x/y \in \mathcal{O}$, and we can suppose by symmetry that the former option holds. Hence $x - y = x(1 - (y/x)) = xt$, with $t := 1 - (y/x) \in \mathcal{O}$. Therefore, in view of what we have just seen, we have $x - y \in \mathcal{P}$.

These verifications show that $\mathcal{P}$ is an ideal. Since $\mathcal{P}$ is by definition the set of non invertible elements of $\mathcal{O}$, we deduce that $\mathcal{P}$ is the unique maximal ideal of $\mathcal{O}$, which is therefore a local ring.

Finally, let $\mathcal{O}$ be a VR of K/k, so $k \subset \mathcal{O}$. Since k is a field, we have that $k^* = k \setminus \{0\} \subset \mathcal{O}^*$ whence $k \cap \mathcal{P} = \{0\}$, which proves the last assertion. $\qquad\qquad\square$

Definition 2.1.2. Let $x \in K$. We say that x *has a* **zero** *at* $\mathcal{P}$ *(or at* $\mathcal{O}$*)* if $x \in \mathcal{P}$ and that *has a* **pole** *at* $\mathcal{P}$ if $x^{-1} \in \mathcal{P}$. Finally, we say that x *is* **regular** *at* $\mathcal{O}$ *if* $x \in \mathcal{O}$.

Remark 2.1.1. This definition already anticipates the idea of viewing the elements of K as functions on Σ_K.

Let $\mathcal{O} \in \Sigma_K$. Let ∞ be a symbol (associated to $\mathcal{O}$) and let us define a map $\varphi : K \to (\mathcal{O}/\mathcal{P}) \cup \{\infty\}$ by sending an $x \in \mathcal{O}$ into its residue class modulo $\mathcal{P}$, and an $x \in K \setminus \mathcal{O}$ (namely an x with a pole at $\mathcal{P}$) to ∞. Observe that if $x \in K \setminus \mathcal{O}$ then $x^{-1} \in \mathcal{O} \setminus \mathcal{O}^* = \mathcal{P}$; this says that φ is a 'homomorphism' (in a wider sense) if in $(\mathcal{O}/\mathcal{P}) \cup \{\infty\}$ we adopt the usual rules $a \pm \infty = \infty$, when $a \in \mathcal{O}/\mathcal{P}$, and $a \cdot \infty = \infty$ if $a \neq 0$, $1/0 = \infty$, $1/\infty = 0$. On the other hand, $\infty \pm \infty$ and $0 \cdot \infty$ are not defined.

Such a map φ is called a **place** of K.

Reciprocally, it is not difficult to verify that a VR $\mathcal{O}$ of K may be obtained starting from a 'place', namely from a 'homomorphism' (in a wider sense) $\varphi : K \to \tilde{K} \cup \infty$, where $\tilde{K}$ is a field; in such a situation one puts $\mathcal{O} := \varphi^{-1}(\tilde{K})$. (Cf. [2] or [17].) For this reason we shall occasionally call $\mathcal{O}$ itself a '**place** of K'.

Proposition 2.1.2. *The set* $\mathcal{P} \subset K$ *determines the VR* $\mathcal{O} \in \Sigma_K$.

Proof. Let $\mathcal{O}, \mathcal{O}' \in \Sigma_K$ have the same maximal ideal $\mathcal{P}$. Let $x \in \mathcal{O} \setminus \mathcal{O}'$. Then $x^{-1} \in \mathcal{O}'$ and actually x^{-1} is in the maximal of $\mathcal{O}'$, which is $\mathcal{P}$ (for otherwise $(x^{-1})^{-1} = x$ would be in $\mathcal{O}'$). But then $1 = x \cdot x^{-1} \in \mathcal{P}$, which is impossible. $\qquad\qquad\square$

Proposition 2.1.3. *If* $\mathcal{O} \in \Sigma_K$, *then* $\mathcal{O}$ *is integrally closed in* K. *In particular, if* $\mathcal{O} \in \Sigma_{K/k}$, $\mathcal{O}$ *contains the algebraic closure of k in K).*

Proof. Let $x \in K$, x integral over $\mathcal{O}$, so there exist $a_1, \ldots, a_n \in \mathcal{O}$ such that $x^n + a_1 x^{n-1} + \ldots + a_n = 0$. If x does not lie in $\mathcal{O}$, we have $x^{-1} \in \mathcal{O}$ by definition. But then the equation $x = -(a_1 + a_2/x + \ldots + a_n/x^{n-1})$ shows that $x \in \mathcal{O}$, absurd. Hence $x \in \mathcal{O}$ in any case, proving the first part. If on the other hand $\mathcal{O}$ is a VR of K/k, then it contains k. Since an element of K which is algebraic over k is automatically integral over k, we get the remaining part of the statement. $\qquad\qquad\square$

An important corollary of this proposition concerns the function fields of one variable, that we have superficially discussed before. We have the:

Proposition 2.1.4. *If K is a function field in one variable over the algebraically closed field k, and if $(\mathcal{O}, \mathcal{P}) \in \Sigma_{K/k}$, then the residue field $\mathcal{O}/\mathcal{P}$ is isomorphic to k under the natural inclusion $k \to \mathcal{O}$.*

Proof. In the sequel we shall tacitly identify k with its isomorphic image in $\mathcal{O}/\mathcal{P}$ (recall Proposition 2.1.1). Suppose now that there exists $x \in \mathcal{O}$ such that the image $\bar{x}$ of x in $\mathcal{O}/\mathcal{P}$ is not in k, and is therefore transcendental over k. Then $f(\bar{x}) \neq 0$ for every nonzero polynomial $f \in k[X]$. Equivalently, $f(x) \notin \mathcal{P}$ for any such polynomial, whence $k(x)$ would be contained in $\mathcal{O}$. But $K/k(x)$ is algebraic, because K/k has transcendence degree 1 by assumption. (Note in fact that x is transcendental over k, since $\bar{x}$ is; or else note that $f(x) = 0$ implies $f(x) \in \mathcal{P}$, which is not the case). Therefore K would be integral over $\mathcal{O}$ and now Proposition 1.3 would lead to $\mathcal{O} = K$, a contradiction. $\qquad\square$

Remark 2.1.2.

  (i) If k is not algebraically closed, a similar argument yields that $\mathcal{O}/\mathcal{P}$ is algebraic over k and in fact a finite extension of k. (See also Theorem 2.5.2 below for another proof.)

 (ii) Note that the conclusion does not hold in general for the function fields of algebraic varieties, namely for the extension fields K/k with finite trascendence degree. (Consider for instance $K = k(X, Y)$ and $\mathcal{O}$ the VR of K consisting of the rational functions whose denominator is not divisible by X.)

Remark 2.1.3 (Valuation rings and geometric points). From Proposition 2.1.4 it follows that *the elements of a function field K/k may be seen as functions with values in* $\mathbb{P}_1(k) = k \cup \{\infty\}$, *whose domain is the set $\Sigma_{K/k}$ of VR of K/k*: if $x \in K$, the value $x(\mathcal{O})$ shall be ∞ if $x \notin \mathcal{O}$, or the class of x in the residue field $\mathcal{O}/\mathcal{P}$ otherwise. (Note that such class coincides with the unique element $a \in k$ such that $x - a \in \mathcal{P}$. In this way the elements of $k \subset K$ give rise to constant functions.) In this way the VR are viewed as *points*.

This viewpoint coincides with the usual one when $K = k(t)$ is the field of rational functions over k (see also Example 2.1.2 and subsequent Remark, where it is shown that the VR correspond precisely to the points of $\mathbb{P}_1(k)$). Even when K is the function field of a curve there is an analogous geometric interpretation of valuation rings as points. To take an instance, suppose $K = k(x, y)$, where $F(x, y) = 0$ is an irreducible equation over k. If $\mathcal{O}$ is a VR of K/k containing x, y, and if $a, b \in k$ are the classes resp. of x, y modulo $\mathcal{P}$, then taking the class of $F(x, y)$ in $\mathcal{O}/\mathcal{P}$ one finds $F(a, b) = F(x(\mathcal{O}), y(\mathcal{O})) = 0$, so that $\mathcal{O}$ 'corresponds' to the point $(a, b) \in k^2$ on the plane curve $F = 0$ associated to K. In the applications of Theorem 2.2.1 we shall see how in general the geometric points always come from suitable VR, in the fashion that we have just illustrated. However, one has to take into account that this correspondence is not bijective for general function fields. It is 'almost' bijective (up to finitely many cases, in a sense to be clarified in the sequel) in the case of curves. See below for more on this.

2.1.1. Some significant examples

Example 2.1.1. $K = \mathbb{Q}$. Let $\mathcal{O} \in \Sigma_{\mathbb{Q}}$. Then $P := \mathcal{P} \cap \mathbb{Z}$ is an ideal of $\mathbb{Z}$, which is prime because $\mathcal{P}$ is prime. If we had $P = \{0\}$, then each nonzero integer would be invertible in $\mathcal{O}$; but then $\mathcal{O}$ would contain $\mathbb{Q}$, a contradiction. Hence P is a maximal ideal of $\mathbb{Z}$, generated by a prime number p: $P = p\mathbb{Z}$. Then the integers coprime with p are invertible in $\mathcal{O}$ and we deduce that $\mathcal{O}$ contains the localization $\mathbb{Z}_P := \{a/b : a \in \mathbb{Z}, b \in \mathbb{Z} \setminus P\}$. Let then $x \in \mathcal{O} \setminus \mathbb{Z}_P$; then there would exist a positive power p^r of p with $p^r x \in \mathbb{Z}_P^* \subset \mathcal{O}^*$, which is inconsistent, because $p^r x \in p^r \mathcal{O} \subset \mathcal{P}$. Therefore $\mathcal{O} = \mathbb{Z}_P$. Conversely, we immediately see that $\mathbb{Z}_P$ is a VR for every prime p. This describes all the possibilities for a VR of $\mathbb{Q}$. Let us observe that the residue field $\mathcal{O}/\mathcal{P}$ is $\mathbb{F}_p$, if $\mathcal{P}$ corresponds to the prime p. The corresponding place sends a rational r in its residue class modulo p if the denominator of r is prime to p and sends r to $\infty(=\infty_p)$ otherwise.

Remark 2.1.4. The rational numbers may be seen as *functions whose domain is* $\mathrm{Spec}(\mathbb{Z})$ (*i.e.* the set of prime ideals of $\mathbb{Z}$). The 'value' of the rational number x at the nonzero prime $p\mathbb{Z}$ shall be the class of x in $\mathbb{F}_p$ if $x \in \mathbb{Z}_{(p)}$ and ∞ otherwise. The value of x on the zero ideal (also called the *generic point* in this context) is just x. Here the target space changes depending on the point, which is an important difference with the case of function fields (see the next example), when such viewpoint gives back the usual notion of function.

Example 2.1.2. $K = k(x)$, where x is transcendental over k. Let $\mathcal{O} \in \Sigma_{K/k}$.[1] Suppose to start with that $x \in \mathcal{O}$, so $\mathcal{O} \supset k[x]$. Arguing as in the case $K = \mathbb{Q}$ (with $k[x]$ in place of $\mathbb{Z}$) one finds that $\mathcal{O}$ is a localized ring

---

[1] The general determination of the VR of $k(x)$ is rather awkward and unnatural to perform without the assumption $\mathcal{O} \supset k$.

$k[x]_P$, where P is a maximal ideal of $k[x]$, therefore associated biject-ively to a monic irreducible polynomial $p(x) \in k[x]$. This yields a first list of VR of $k(x)/k$.

If $\mathcal{O}$ is associated to the poynomial $p(x)$, the residue field is $k[x]_P/(p(x)) \cong k[x]/(p(x))$, which is an algebraic extension of k, of degree deg p (see Proposition 2.1.4). If ρ is a root of $p(x)$ in some ex-tension field of k, the residue field is isomorphic to $k(\rho)$ (over k); the associated place $\varphi = \varphi_P$ sends $r(x)$ to $r(\rho)$ ($= \infty$ if ρ is a pole of r).

Suppose now that $\mathcal{O}$ does not contain x. Then $y := x^{-1} \in \mathcal{O}$ and actually $y \in \mathcal{P}$ (since $y^{-1} = x \notin \mathcal{O}$). On the other hand (as we have seen before) $\mathcal{P} \cap k[y]$ is a maximal ideal of $k[y]$, necessarily generated by y and $\mathcal{O}$ coincides with the localized ring $k[y]_{(y)}$. Such a ring consists of the rational functions of the shape $r(y) = a(y)/b(y)$ where a, b are polynomials and $b(0) \neq 0$. The residue field is now $k[y]/(y) \cong k$.

To read this in terms of x, let us put $d := \max(\deg a, \deg b)$; then we may write $r(y) = a(x^{-1})/b(x^{-1}) = A(x)/B(x)$ where $A(x) = x^d a(x^{-1})$ and $B(x) = x^d b(x^{-1})$ are polynomials; let us observe that since $b(0) \neq 0$, deg $B = d$, whereas in any case deg $A \leq d$. In conclu-sion, $\mathcal{O}$ consists of the rational functions *without pole at infinity*, namely of the shape $A(x)/B(x)$ where deg $B \geq$ deg A. The associated place sends a function to its value at ∞. This single VR, added to the preced-ing list, completes the classification.

Remark 2.1.5.

(i) Suppose for example that k is algebraically closed. Then the max-imal ideals of $k[x]$ are bijectively associated to the polynomials $x - \alpha$, $\alpha \in k$, and hence to the elements $\alpha \in k$. Hence the clas-sification shows that the corresponding VR correspond to the points of k (in the first list) plus the VR associated to ∞. In other words, the VR correspond to points of $\mathbf{P}_1(k)$. A place $\varphi = \varphi_Q$ associated to a point $Q \in \mathbb{P}_1$ sends the element $r(x) \in k(x)$ to $\varphi(r) = r(Q)$, where $r(Q)$ is to be read as ∞ if Q is a pole of r. (This illustrates in a precise way the comments to Proposition 2.1.4, where the VR were seen as points.)

   The finite extensions of $\mathbb{Q}$ or $k(x)$ shall be considered in the sequel (the corresponding VR extend those of the base field).

(ii) For $K = k(x_1, \ldots, x_n)$, with algebraically independent x_i over k, the classification is much more difficult already for $n = 2$; it has been given by Zariski (see [8], also for references). To our know-ledge, it is not been formulated for $n > 2$.

Example 2.1.3. : $K = k((x))$. This field (which is the fraction field of the formal power series ring $k[[x]]$, and is called the field of 'Laurent series') has infinite transcendence degree over k, which renders practically impossible a useful general classification of the VRs of K/k. On the other hand, there exists a single VR containing $k[[x]]$. In fact, $k[[x]]$ has a unique prime ideal $P \neq \{0\}$, generated by x, and this ideal is actually maximal. The VR is, as above, the localized ring, which now is the whole ring: $\mathcal{O} = k[[x]]_{(x)} = k[[x]]$, and the corresponding place assigns to a series in $k((x))$ its value at 0 (value which is ∞ if the series does not lie in $k[[x]]$).

Exercise 2.1.1. Let $K = k(x, y)$ be the field of rational functions in the independent indeterminates x, y (char $k = 0$). Let $s(x)$ be the formal series $s(x) = \exp(x) - 1 = x + \frac{x^2}{2} + \frac{x^3}{6} + \ldots \in k[[x]]$. Prove that if $r(x, y) \in K$ then $r(x, s(x))$ is defined (as a formal series in $k((x))$).

Let now $\mathcal{O} \subset K$ be the set of rational functions $r(x, y)$ such that $r(x, s(x))$ has not a pole at $x = 0$. Prove that $\mathcal{O}$ is a VR of K. Let further $\mathcal{O}'$ be the VR of K obtained in the same way, but exchanging the roles of x, y. Prove that $\mathcal{O} \neq \mathcal{O}'$.

Example 2.1.4. Let $\mathcal{D}$ be an open connected nonempty subset of $\mathbb{C}$ and let K be the field of meromorphic functions on $\mathcal{D}$. Further, let $z_0 \in \mathcal{D}$. Then the set $\mathcal{O} = \mathcal{O}_{z_0}$ of the functions in K which are regular at z_0 is a VR of $K/\mathbb{C}$. In fact, $\mathcal{O}$ is a ring $\neq K$ and if a function $f \in K$ is not in $\mathcal{O}$, then it has a pole at z_0 and hence $1/f \in \mathcal{O}$. The maximal ideal $\mathcal{P}$ is made up of the functions in K which vanish at z_0 and $\mathcal{O}/\mathcal{P} \cong \mathbb{C}$. (In this situation it is not true that all VRs of K are of the above type; see Exercise 2.2.1 (ii) for an example.)

2.2. Existence and extensions of valuation rings

The following result will be crucial in what follows.

Theorem 2.2.1. *Let K be a field, $A \subset K$ be a subring and let $I \neq \{0\}$ be a proper ideal of A. Then there exists a VR $(\mathcal{O}, \mathcal{P})$ of K such that $A \subset \mathcal{O}$ and $I \subset \mathcal{P}$.*

Proof. Let us order by inclusion the rings $B \supset A$ in K such that $IB \neq B$ (observing that A is among them). Every chain (*i.e.* a totally ordered subset) has a maximal element, given by the union: indeed, the union U of the rings of the chain is a subring of K containing A; if we had $IU = U$, then 1 would be of the shape $\sum_{i=1}^{n} \omega_i u_i$, $\omega_i \in I$, $u_i \in U$. But then, taking as B a ring in the union containing all the u_i we would have $IB = B$, a contradiction.

Therefore, by Zorn's lemma, there exists a such a ring $\mathcal{O}$ maximal under inclusion; it contains A and let us now show that it has the other properties in the statement.

Let $x \in \mathcal{O}, x \equiv 1 \pmod{I\mathcal{O}}$. We claim that the ring $\tilde{\mathcal{O}} := \mathcal{O}[x^{-1}]$ is such that $I\tilde{\mathcal{O}} \neq \tilde{\mathcal{O}}$: indeed, otherwise we would have an equality $1 = \sum_{i=1}^{n} \omega_i u_i x^{-m}$ for elements $\omega_i \in I, u_i \in \mathcal{O}$ and for a large enough m. Then $x^m = \sum_{i=1}^{n} \omega_i u_i$ and therefore $1 = (1 - x^m) + \sum_{i=1}^{n} \omega_i u_i \in I\mathcal{O}$, against the assumption that $I\mathcal{O} \neq \mathcal{O}$.

Then, by maximality we have $\tilde{\mathcal{O}} = \mathcal{O}$ and hence x is invertible in $\mathcal{O}$.

Let now $u \in K \setminus \mathcal{O}$, so that $\mathcal{O}[u]$ contains $\mathcal{O}$ properly; by maximality we have $1 \in I\mathcal{O}[u]$, i.e., $1 = \sum_{i=0}^{m} \rho_i u^i$, for suitable $\rho_i \in I\mathcal{O}$. We can further assume that m (which must be > 0) is minimal for expressions of this type. Setting $x = 1 - \rho_0$ we have $x \equiv 1 \pmod{I\mathcal{O}}$; then x is invertible in $\mathcal{O}$ and we have then an equality $1 = \sum_{i=1}^{m} (\rho_i x^{-1}) u^i = \sum_{i=1}^{m} v_i u^i$, where also the $v_i := \rho_i x^{-1}$ lie in $I\mathcal{O}$.

We want to show that $u^{-1} \in \mathcal{O}$: if this does not hold, as before we would obtain elements $\sigma_j \in I\mathcal{O}$ such that $1 = \sum_{j=1}^{n} \sigma_j u^{-j}$, where again we can assume that n is minimal. Let us show that these last two expressions (for u and for u^{-1}) are incompatibile, assuming by symmetry that $m \geq n$. Multiplying the second equality by u^m we obtain $u^m = \sum_{j=1}^{n} \sigma_j u^{m-j}$ and hence, substituting for u^m in the former one, we get $1 = \sum_{i=1}^{m-1} v_i u^i + \sum_{j=1}^{n} v_m \sigma_j u^{m-j}$. However this is an equation of the first type, where the maximal exponent of u is decreased, which yields the required contradiction by minimality of m.

Summing up, we have shown that either $u \in \mathcal{O}$ or $u^{-1} \in \mathcal{O}$. Since $I\mathcal{O} \neq \mathcal{O}$ and since I is nonzero, $\mathcal{O}$ is strictly contained in K (because $IK = K$) and hence is a VR of K containing A. Finally, if there exists $x \in I \setminus \mathcal{P}$, x would be invertible in $\mathcal{O}$ and therefore $I\mathcal{O}$ would contain $x \cdot x^{-1} = 1$, a contradiction which proves that $I \subset \mathcal{P}$, and the theorem. $\square$

Remark 2.2.1.

(i) The condition that the ideal I in the theorem is nonzero cannot be omitted, as is illustrated by the choice $A = K$.

(ii) The last (crucial) part of the proof may seem a bit artificial; it is inspired by the two basic relations for u, u^{-1}, which show respectively that u^{-1} and u are integral over $I\mathcal{O}$. If we would like to use the fact (not proved here) that such integral elements form a subring, we could obtain at once the contradiction $1 = u \cdot u^{-1} \in I\mathcal{O}$, simplifying a bit the argument.

Exercise 2.2.1.

(i) Let x, y be algebraically independent over the field k. Prove that there exist infinitely many VRs $(\mathcal{O}, \mathcal{P})$ of $k(x, y)/k$ such that $x, y \in \mathcal{P}$.

(Hint: let $t := x/y$; consider VRs containing $k[x, t]$ and such that $x, t - a \in \mathcal{P}$ for a certain $a \in k$.)

(ii) Notation as in Example 2.1.4, with $\mathcal{D} = \mathbb{C}$, prove that there exist VRs of K distinct from $\mathcal{O}_{z_0}$, no matter $z_0 \in \mathbb{C}$.

(Hint: let for instance A be the ring of integral functions and I be the ideal of those which vanish on all but finitely many integers - *e.g.* $\sin 2\pi z/(z - m)$. Applying Theorem 2.2.1 one obtains the sought example.)

2.2.1. Some applications of Theorem 2.2.1

Let us go ahead by showing some significant applications of Theorem 2.2.1. They have independent interest, however the most fundamental (for us) applications to the case of extensions of VRs of $\mathbb{Q}$ or $k(x)$ shall appear later.

Application A. *Let* $K = k(x_1, \ldots, x_r)$ $(r \geq 1)$ *be an extension of k, proper* (i.e. $\neq k$) *and finitely generated and let J be the ideal determined by* $(x_1, \ldots, x_r)$ *in* $k[X_1, \ldots, X_r]$. *Then, if* $(\xi_1, \ldots, \xi_r) \in k^r$ *is a zero of J, there exists a VR $(\mathcal{O}, \mathcal{P})$ of K/k with $x_i - \xi_i$ in $\mathcal{P}$.*

Preliminary to the proof of this statement, recall that here J is the ideal consisting of the $f(\mathbf{X})$ such that $f(\mathbf{x}) = 0$, *i.e.* J is the kernel of the homomorphism $\phi : k[X_1, \ldots, X_n] \to k[x_1, \ldots, x_n]$ mapping X_i to x_i. A *zero* of J is by definition a common zero of all the polynomials in J.

As to the proof, let us apply Theorem 2.2.1 on taking $A = k[x_1, \ldots, x_r]$, $I = \{f(x_1, \ldots, x_r) : f \in k[X_1, \ldots, X_n], f(\xi_1, \ldots, \xi_r) = 0\}$.

The assumptions are verified: surely I is an ideal (it is $\phi(J)$ in the above notation), contains $x_i - \xi_i, i = 1, \ldots, r$, and in particular is nonzero (because $A \neq k$). If we had $I = A$, then an equality $1 = f(x_1, \ldots, x_r)$ would hold with f a polynomial vanishing at $(\xi_1, \ldots, \xi_r)$. Now, the equality shows that $f(X_1, \ldots, X_r) - 1 \in J$; but then, since $(\xi_1, \ldots, \xi_r)$ is a zero of J, we would obtain $f(\xi_1, \ldots, \xi_r) - 1 = 0$, a contradiction.

Let then $(\mathcal{O}, \mathcal{P})$ be a VR of K as in the conclusion of the theorem, namely containing A and such such that $I \subset \mathcal{P}$; then, in view of the fact that $x_i - \xi_i \in I$, we have the assertion.

Application B. *Let* $K = k(x, y)$ *be a function field in one variable over k, and let us suppose that $F(x, y) = 0$ for some irreducible polynomial $F \in k[X, Y]$. Then, if $F(a, b) = 0$ for some $(a, b) \in k^2$, there exists a VR $(\mathcal{O}, \mathcal{P})$ of K/k with $x - a, y - b \in \mathcal{P}$.*

This statement is an easy corollary of A. Indeed, let $J \subset k[X, Y]$ be the ideal determined by (x, y). Since F is irreducible and since x or y is transcendental over k by assumption, J must be generated by F (see the previous Chapter) and hence (a, b) is a zero of J, which yields the assertion.

Remark 2.2.2. In these two statements we see how the *geometric points* $(\xi_1, \ldots, \xi_r)$ (in the first case) and (a, b) (in the second one) correspond to suitable VR. (This correspondence meaning that $\mathcal{P} \cap k[x_1, \ldots, x_r]$ is the maximal ideal generated by the $x_i - \xi_i$, associated to $(\xi_1, \ldots, \xi_r)$; see also previous related comments and exercises.)

One should also recall that these VRs are by no means uniquely determined by the point (see *e.g.* the last exercise, part (i))). However in the case of curves and their function fields this correspondence is essentially bijective (*i.e.*, with at most finitely many exceptions).

Application C. *Let K/k be a field extension and let $x \in K$ be a transcendental element over k. Then x admits at least one zero and at least one pole (in suitable VRs of K/k).*

For a proof, let us set in Theorem 2.2.1 $A = k[x]$, $I = xA$. Since x is transcendental over k, I is nonzero and proper. Then the conclusion of the theorem ensures the existence of a VR $(\mathcal{O}, \mathcal{P})$ of K containing A and such that $x \in \mathcal{P}$, which says that x has a zero at $\mathcal{O}$ (cf. Def. 2.1.2). Considering $1/x$ in place of x we obtain the remaining part.

Application D. *If K is a transcendental extension of k, there exist infinitely many VRs of K/k.*

In fact, let $x \in K$ be transcendental over k; then there exist infinitely many VRs $(\mathcal{O}, \mathcal{P})$ of $k(x)/k$. (There exist one such VR for each monic irreducible polynomial; there exist infinitely many such polynomials, because an algebraic closure of a field cannot be a finite field.) It suffices now to extend to K each of these VRs, taking for instance $A = k[x]$, $I = \mathcal{P} \cap A$ in Theorem 2.2.1.

Application E. *Let A be a subring of a field K. Then the integral closure A in K is the intersection of the VRs of K containing A.*

For a proof, let B be the integral closure in question (*i.e.*, the set of elements of K which are integral over A). Let $x \in B$ and let $\mathcal{O}$ be a VR of K containing A; since x is integral over A, it is *a fortiori* integral over $\mathcal{O}$ and hence lies in $\mathcal{O}$ (by Proposition 2.1.3). This proves half of the statement.

Conversely, let $x \in K \setminus B$. Then $y := x^{-1}$ is not invertible in $A[y]$ (for otherwise we would have $x = y^{-1} \in A[y] = A[x^{-1}]$, which gives an integral equation for x over A). Therefore y lies in a maximal ideal I of $A[y]$. Certainly I is nonzero (it contains y) and hence we can apply Theorem 2.2.1 (with $A[y]$ in place of A), obtaining the existence of a VR $(\mathcal{O}, \mathcal{P})$ of K with $A[y] \subset \mathcal{O}$ and $I \subset \mathcal{P}$. If we had $x \in \mathcal{O}$, we would also have $1 = yx \in I\mathcal{O} \subset \mathcal{P}$, a contradiction. Hence $x \notin \mathcal{O}$ and x does not lie in the intersection in question, concluding the argument.

Remark 2.2.3. Note that this result directly implies that the integral closure of a subring of a field is itself a subring.

Example 2.2.1. The ring $\mathbb{Z}$ is the intersection of the VRs of $\mathbb{Q}$ (see Example 2.1.1) and in fact $\mathbb{Z}$ is integrally closed (as follows easily from unique factorization). Similarly, the ring of algebraic integers (that is the algebraic numbers integral over $\mathbb{Z}$) in a number field K^2 coincides with the intersection of the VRs of K.

Further examples occur within function fields; for instance, the polynomial ring $k[x]$ is the intersection of those VRs of $k(x)/k$ which contain x. (Here everything depends on the choice, not canonical, of the generator x, contrary to the case of $\mathbb{Q}$.)

Remark 2.2.4. *If A is local with maximal ideal M, the conclusion of Application E holds even if we limit the intersection to those VRs $(\mathcal{O}, \mathcal{P})$ of K with $A \subset \mathcal{O}$ and $M \subset \mathcal{P}$.*

The first part of the proof remains unchanged. As to second part, one may proceed similarly: one start by observing that if $x \in K \setminus B$ then $y = x^{-1}$ is not invertible in $A[y]/M[y]$. (Otherwise we would have $1 = m(y) + l(y)y, m \in M[y], l(y) \in A[y]$, whence $1 = m_0 + l_1(y)y$, $m_0 \in M, l_1(y) \in A[y]$; but $1 - m_0 \in A^*$ and hence y would be invertible in $A[y]$, and as above x would be integral over A, a contradiction.) Then $M[y] + yA[y]$ is contained in a maximal ideal I of $A[y]$ and now the proof may be concluded exactly as above.

Application F. *If a finitely generated algebra K/k is a field, then it is a finite extension field.*

The statement is one of the several equivalent versions of the celebrated *Nullstellensatz* of Hilbert. See Appendix A below for a discussion and for other approaches to a proof of Hilbert's theorem.

To prove the above claim, write $K := k[x_1, \ldots, x_m]$. If the conclusion does not hold, let $z_1, \ldots, z_r$ be a transcendence basis for K/k, with

---

[2] By definition this is a finite extension of $\mathbb{Q}$.

$r > 0$. Each one among $x_1, \ldots, x_m$ then satisfies a non-trivial algebraic equation with coefficients in $A := k[z_1, \ldots, z_r]$. Let $f(z_1, \ldots, z_r) \in A \setminus \{0\}$ be the product of the leading coefficients of such m polynomials. Let then I be a nonzero maximal ideal of A, not containing f. (Such an ideal exists, and it is here that we use the condition $r > 0$: it suffices to take a maximal ideal of A containing $z_1 f + 1$, which has positive degree and is therefore not invertible. Alternatively, if $\xi_1, \ldots, \xi_r$ are elements in an algebraic extension k' of k and such that $f(\xi_1, \ldots, \xi_r) \neq 0$, it suffices to take as I the set of elements of A which vanish at $(\xi_1, \ldots, \xi_r)$; I does not contain f, it is nonzero - because $r > 0$ and the ξ_i are algebraic over k - and it is maximal because A/I is isomorphic to $k[\xi_1, \ldots, \xi_r] = k(\xi_1, \ldots, \xi_r)$.)

Now let, as in Theorem 2.2.1, $(\mathcal{O}, \mathcal{P})$ be a VR of K with $A \subset \mathcal{O}$ and $I \subset \mathcal{P}$. Since I does not contain f, we see that also $\mathcal{P}$ does not contain f (for otherwise $\mathcal{P}$ would contain $I + (f)$, which equals A since I is maximal). Observe that, for $i = 1, \ldots, m$, x_i satisfies an equation over A with leading coefficient f. Therefore, since f is invertible in $\mathcal{O}$ (because f is not in $\mathcal{P}$, as we have seen), each x_i is integral over $\mathcal{O}$. But $\mathcal{O}$ is integrally closed (see Proposition 2.1.3) and hence $x_i \in \mathcal{O}$ for every $i = 1, \ldots, m$, hence $\mathcal{O}$ contains $k[x_1, \ldots, x_m] = K$, a contradiction which finally proves the claim.

Exercise 2.2.2. Let A be a domain and write a polynomial $f \in A[X]$ as $f(X) = a \prod_{i=1}^{d}(X - \rho_i)$, where the ρ_i lie in a field K containing A. Prove that for all subsets S of $\{1, \ldots, d\}$, $a \prod_{i \in S} \rho_i$ is integral over A. (Hint: Let $\mathcal{O}$ be a VR of K containing A. If $\rho_1, \ldots, \rho_s \notin \mathcal{O}$ and $\rho_i \in \mathcal{O}$ for $i > s$, prove that $a \prod_{i=1}^{s}(X - \rho_i) \in \mathcal{O}[X]$.)

A more explicit proof of this result is due to Kronecker (see [31]).

We conclude this paragraph with two further useful results on the extension of VRs.

Proposition 2.2.2. *If L/K is a field extension and $(\mathcal{O}, \mathcal{P})$ is a VR of L not containing K, then $(\mathcal{O} \cap K, \mathcal{P} \cap K)$ is a VR of K.*

Proof. Let $A := \mathcal{O} \cap K$; if $x \in K$, then either $x \in \mathcal{O}$, and hence $x \in A$, or $x^{-1} \in \mathcal{O} \cap K = A$. This shows that A is a VR of K. If now $x \in A \setminus A^*$, we have either $x = 0$ or $x^{-1} \notin \mathcal{O}$, which implies that $x \in \mathcal{P}$. Therefore $A \setminus A^* \subset \mathcal{P}$. Finally, if $x \in \mathcal{P} \cap K$, then $x \in A$; on the other hand x cannot lie in A^*, because x^{-1} is not in $\mathcal{O}$. This proves the opposite implication. $\square$

Proposition 2.2.3. *Let L/K be a field extension and let $(\mathcal{O}, \mathcal{P})$, $(\mathcal{O}', \mathcal{P}')$ be VR in K, L respectively, such that $\mathcal{O} \subset \mathcal{O}'$ and $\mathcal{P} \subset \mathcal{P}'$. Then $\mathcal{O} = \mathcal{O}' \cap K$.*

Proof. The inclusion $\mathcal{O} \subset \mathcal{O}' \cap K$ is clear. Let now $x \in \mathcal{O}' \cap K$, $x \notin \mathcal{O}$. Then $y := x^{-1} \in \mathcal{P}$ and hence $y \in \mathcal{P}'$, against the fact that $x = y^{-1}$ lies in $\mathcal{O}'$. We then have a contradiction which proves the opposite inclusion and the claim. $\qquad\square$

Remark 2.2.5. Observe that the assumption $\mathcal{O} \subset \mathcal{O}'$ implies $\mathcal{P}' \cap \mathcal{O} \subset \mathcal{P}$; however that $\mathcal{P} \subset \mathcal{P}'$ is not automatic: for instance, let $K = \mathbb{Q}$, $L = \mathbb{Q}(x)$, and let $\mathcal{O}$ be any VR of $\mathbb{Q}$, $\mathcal{O}'$ be the VR corresponding to the polynomial x. We have $\mathcal{P}' \cap \mathbb{Q} = \{0\}$.

2.3. Discrete valuation rings

Definition 2.3.1. A *discrete valuation ring* (abbr. DVR) is a principal ideal domain with a single nonzero prime ideal $\mathcal{P}$. A *local parameter* is any generator of $\mathcal{P}$.

We also say that $\mathcal{O}$ is a DVR of the field K if $\mathcal{O}$ is a DVR whose fraction field is K.

We also note that a principal VR is necessarily a DVR (compare next exercise).

Remark 2.3.1. The terminology 'local parameter' arises from most significant instances, such as in the examples and exercises below.

Sometimes we shall refer to a DVR $(\mathcal{O}, \mathcal{P})$ of K as a '**prime**' or '**point**' of K. Such terminology arises *e.g.* from the cases of $\mathbb{Z}$ or $k(x)$. (See exercises below.)

Proposition 2.3.1. *Let $\mathcal{O}$ be a DVR, t a local parameter, $\mathcal{P} = t\mathcal{O}$ its nonzero prime ideal. Then we have $\mathcal{P}^n = t^n \mathcal{O}$ and $\cap_{n \geq 0} t^n \mathcal{O} = \{0\}$.*

Proof. The equality $\mathcal{P}^n = t^n \mathcal{O}$ is clear. The intersection in question is an ideal I of $\mathcal{O}$, hence principal, say generated by u. Since $u \in t^n \mathcal{O}$, we may write $u = t^n v_n$ for suitable $v_n \in \mathcal{O}$. Then (for every $n \geq 1$), $t v_1 = t^n v_n$, whence ($\mathcal{O}$ is a domain and $t \neq 0$) $v_1 = t^{n-1} v_n$, which shows that $v_1 \in I$. Hence $v_1 = vu$ (with $v \in \mathcal{O}$) and $u = tvu$. If we had $u \neq 0$, we would obtain $1 = tv$, which is inconsistent because t generates a proper ideal. Hence $u = 0$ as wanted. $\qquad\square$

This proposition (which is a special case of a result by Krull, see [2]) shows that if $a \in \mathcal{O}, a \neq 0$, there exists a maximal integer $n \geq 0$ such that $a \in \mathcal{P}^n = t^n\mathcal{O}$; in particular, such integer does not depend on the generator t. Let us put

$$v(a) = v_\mathcal{P}(a) = n;$$

by convention let us also put $v(0) = \infty$.

Proposition 2.3.2. *If* $v(a) = n \in \mathbb{N}$, *we can write* $a = t^n q$, *where* $q \in \mathcal{O} \setminus \mathcal{P} = \mathcal{O}^*$. *Moreover,* $(\mathcal{O}, \mathcal{P})$ *is a VR of the fraction field of* $\mathcal{O}$.

Proof. If $v(a) = n$ we have $a \in \mathcal{P}^n \setminus \mathcal{P}^{n+1}$, hence $a = t^n q$ where $q \in \mathcal{O} \setminus \mathcal{P}$. But $\mathcal{P}$ is the unique nonzero prime ideal of $\mathcal{O}$, and hence $\mathcal{O}$ is local, $\mathcal{O} \setminus \mathcal{P} = \mathcal{O}^*$ and $q \in \mathcal{O}^*$. Let further $x = a/b$ be an element of the fraction field of $\mathcal{O}$, where $a, b \in \mathcal{O}, b \neq 0$. If $x \notin \mathcal{O}$ then $x \neq 0$ and we can write $a = t^m a', b = t^n b'$, with $v(a) = m, v(b) = n$ and t a local parameter. By Proposition 2.3.1, a' and b' lie in $\mathcal{O}^*$, so that $x = t^{m-n} u$ with $u \in \mathcal{O}^*$. Since $x \notin \mathcal{O}$, we have $m < n$; but then $x^{-1} = t^{n-m} u^{-1} \in \mathcal{P} \subset \mathcal{O}$. This shows that $\mathcal{O}$ is a DVR of its fraction field. $\square$

Proposition 2.3.3. *The function* $v = v_\mathcal{P} : \mathcal{O} \to \mathbb{N} \cup \{\infty\}$ *is finite on* $\mathcal{O} \setminus \{0\}$, *not constant on* $\mathcal{P}$ *and satisfies:*

$$v(ab) = v(a) + v(b), \tag{2.1}$$

$$v(a + b) \geq \min\{v(a), v(b)\}, \tag{2.2}$$

with equality if $v(a) \neq v(b)$.

Proof. For a local parameter t we have $v(t^n) = n$, whence $v(\mathcal{P}) = \mathbb{N}^+ \cup \{\infty\}$. By Proposition 2.3.1, we can have $v(x) = \infty$ only if $x = 0$.

Now, if $ab = 0$ the assertions are obvious (with the rule $\infty + h = \infty$ for $h \in \mathbb{Z}$). Otherwise, putting $v(a) = m, v(b) = n$, we can write $a = t^m a', b = t^n b'$, with $a', b' \in \mathcal{O}^*$. Then $ab = t^{m+n} a' b'$, proving (2.1), since $a' b' \in \mathcal{O}^*$.

Let now for instance $m \leq n$; then $a + b = t^m(a' + t^{n-m} b')$ which proves (2.2), since $a' + t^{n-m} b' \in \mathcal{O}$. If on the other hand $m < n$, we have $t^{n-m} b' \in \mathcal{P}$, so $a' + t^{n-m} b' \in \mathcal{O} \setminus \mathcal{P} = \mathcal{O}^*$ (because $a' \notin \mathcal{P}$), proving even the last assertion. $\square$

We now see that the domain of the function $v = v_\mathcal{P} : \mathcal{O} \to \mathbb{N} \cup \{\infty\}$ can be extended to the fraction field K of $\mathcal{O}$, setting

$$v(a/b) = v(a) - v(b), \qquad a, b \in \mathcal{O}, \quad ab \neq 0.$$

Indeed, Proposition 2.3.3 shows that the definition is consistent and that, moreover, the properties (2.1), (2.2) still hold for $a, b \in K$. Further, by induction or otherwise one can verify a generalization of (2.2):

$$v(a_1 + \ldots + a_n) \geq \min_{i=1,\ldots,n} \{v(a_i)\}, \tag{2.3}$$

with equality if the minimum is attained for a unique index i.

Definition 2.3.2. The (surjective) function $v = v_{\mathcal{P}} : K \to \mathbb{Z} \cup \{\infty\}$ is called **order function** and $v(a)$ is called **order** of a (with respect to $\mathcal{P}$). We say that v (or $\mathcal{O}$, or also $\mathcal{P}$) is a **zero** (resp. **pole**) for a if $v(a) > 0$ (resp. $v(a) < 0$).

To an order function we may associate a non-archimedean **absolute value**[3] of the fraction field K of $\mathcal{O}$, as follows: For a given positive real number $c < 1$ one sets:

$$|x| = |x|_v := c^{v(x)}, \qquad x \in K, \tag{2.4}$$

with the convention $c^\infty = 0$. Since $v(K^*) = \mathbb{Z}$, The above propositions yield:

Proposition 2.3.4. *We have $|x| \in \{c^m : m \in \mathbb{Z}\} \cup \{0\}$ if $x \in K$, and $|x| = 0$ if and only if $x = 0$. Also, we have $|xy| = |x||y|$ for $x, y \in K$ and, for $x_1, \ldots, x_n \in K$, it holds that $|x_1 + \ldots + x_n| \leq \max_{i=1,\ldots,n} |x_i|$ with equality if the maximum is attained at a single index.*

These properties say in particular that the absolute value $| \cdot |$ induces a distance $d(\cdot, \cdot)$ on K^2, defined by

$$d(x, y) := |x - y|. \tag{2.5}$$

Such a distance in turn gives a topology on the field K, which becomes a **metric space**. It is an easy matter to verify that the field operations are continuous, which makes K a topological field.

By definition of this topology the function $| \cdot | : K \to \mathbb{R}$ is continuous. Also, by the above the set of values has 0 as the only limit point; it follows that $| \cdot |$ is locally constant as a function on K^*. This is a most important property, which in fact holds for all ultrametric absolute values (independently of the discreteness of the set of values on K^*), as is easy to see.

---

[3] Or **valuation**.

Remark 2.3.2. The choice of c is unimportant when considering a single DVR (see Remark(ii) below), but may become relevant 'globally', that is when all DVR of a field are considered simultaneously. (Indeed, for example, on number fields and function fields of one variable, a *product formula* holds which involves all the absolute values, provided they are suitably normalized. See for instance [16].)

By (2.2), the distance $d(\cdot, \cdot)$ verifies a strong form of the triangle inequality. When such inequality holds, both the absolute value and the associated distance are said **ultrametric**.

Note that (setting $x = y = 1$), the above properties imply $|1| = 1$, whereas $|n| \leq 1$ for $n \in \mathbb{Z}$.[4]

We also observe that the existence of $x \in K^*$ such that $0 < |x| \neq 1$ (*e.g.* x a local parameter) ensures that the mentioned topology is non-discrete (since $x^n \to 0$ for $n \to +\infty$).

We shall often speak of '**place** v' to denote indifferently either the absolute value (up to the choice of c) or the associated order function v, or the DVR, or the place associated to it.

Remark 2.3.3.

(i) It is readily seen that, conversely, if $|\cdot| : K \to \mathbb{R}_{\geq 0}$ is a function vanishing only at 0, not constantly equal to 1 on K^* and such that equations (2.1), (2.2) hold, then the set $\mathcal{O} := \{x \in K : |x| \leq 1\}$ is a VR of K.

(ii) We remark that different choices of c $(0 < c < 1)$ change the absolute value and distance but do not change the associated topology, as the interested reader will very easily verify. Actually, it is not too difficult to prove that two valuations $|\cdot|_1, |\cdot|_2$ on a field K induce the same topology if and only if $|\cdot|_1 = |\cdot|_2^l$ for some positive number l. (See, *e.g.*, [10], or the hinted Exercise 2.3.1(x) below.)

(iii) We note that a field K equipped with the topology coming from an ultrametric absolute value $|\cdot|$ is totally disconnected. In fact, let x, y be distinct points in K, so $d := |x - y| > 0$. Now observe that K is the union of the disjoint nonempty sets $\{z \in K : |z - x| < d\}, \{z \in K : |z - x| \geq d\}$, which contain respectively x, y. That the first set is open is clear; and the second one also is, by

---

[4] Recall the *principle of Archimedes*, saying that '$|n|$ is unbounded for $n \in \mathbb{N}$': this is not verified for this type of absolute value. Conversely, it is not difficult to show that if an absolute value is bounded on $\mathbb{N}$ then it is ultrametric.

the above mentioned locally-constancy property, which implies that 'open' balls are closed.

(iv) These facts make the theory of analytic continuation on complete fields equipped with ultrametric valuations (see below) far different from and more delicate than the complex case. For instance there exist analytic locally constant but nonconstant functions. All of this led to the creation of several new concepts, *e.g.*, the so-called *rigid analytic geometry*. (See *e.g.* [10] for some fundamentals and for references.)[5]

Example 2.3.1.

(i) Case $K = \mathbb{Q}$: We have seen that all the VRs of $\mathbb{Q}$ are localizations of the type $\mathbb{Z}_{(p)}$ where (p) is the prime ideal of $\mathbb{Z}$ generated by the prime number p. It is very easy to see that all these rings are principal and in fact DVR. The order function associated to p sends an integer $a \in \mathbb{Z}$ to the maximum power of p dividing a.

Usually, the corresponding absolute value is denoted by $| \cdot |_p$ and is defined by choosing $c = 1/p$; it is also called p-**adic absolute value**. In this topology the 'small' rational numbers are those whose numerator is divisible by a 'large' power of p.

The usual absolute value on $\mathbb{Q}$ is not ultrametric and does not correspond to a VR. In a sense one could say it is analogue to the VR of $k(x)$ corresponding to the *point at infinity* (which is the unique VR for which x has a pole, and which therefore depends on the choice for x, unlike the case of $\mathbb{Q}$).

We shall study later the case of finite extensions of $\mathbb{Q}$, where several results remain true.

(ii) Case $K = k(x)$: This case is completely analogous to the case of $\mathbb{Q}$, in view of Example 2.1.2. All the VRs are DVRs; those which contain $k[x]$ are associated to maximal ideals, generated by irreducible polynomials. When k is algebrically closed, the VRs are associated to the points of $\mathbb{P}_1(k)$; the order function associated to a point $\alpha \in k \cup \{\infty\}$ assigns to a rational function its **multiplicity** of vanishing at α (which shall be a negative integer in case α is a pole). We shall see later that, as for the case of $\mathbb{Q}$, the situation is similar to the case of finite extensions of $k(x)$.

---

[5] More recently, a theory of 'Berkovitch spaces' has been developed in this direction, with advantages over previous concepts; see for instance [4].

In the case of several variables (*e.g.*, $K = k(x, y)$) not all the VRs are DVRs (see the exercises below); but see (v) below for important examples of DVRs for this case.

(iii) Case $K = k((x))$: Example 2.1.3 produces a VR of this field, which is also a DVR. The order function sends a formal series to its multiplicity at the origin. As we shall see, $k((x))$ may be viewed as the *completion* of $k(x)$ with respect to the distance induced by the the DVR corresponding to the point 0.

(iv) Case of the Example 2.1.4 (*i.e.* analytic functions at a point): In this case too, the ring $\mathcal{O} = \mathcal{O}_{z_0}$ defined therein is a DVR; the order function is the multiplicity at the point z_0. A local parameter establishes a biholomorphic correspondence between a neighborhood of z_0 in $\mathcal{D}$ and an open disk around 0 in $\mathbb{C}$.

Contrary to the case of general VR, it may be proved that every DVR of the field of meromorphic functions on $\mathbb{C}$ is of this type. (This is due to Hironaka; see [17] Ex. 20, Ch. XII.)

(v) In analogy to the case $K = k(x)$ of Example 2.1.2 we have the higher dimensional case $K = k(x_1, \ldots, x_n)$ where the x_i are algebraically independent over k. We may define a DVR starting from any irreducible polynomial $P \in k[x_1, \ldots, x_n]$ and defining the order function as above. Now not all the VRs are DVRs. (See *e.g.*, Exercise 2.2.1(i) and the coming exercises.)

One has natural DVRs also in the case of function fields of normal varieties of arbitrary dimension (the previous case being associated to $\mathbb{P}_n$). They correspond to *prime divisors*, *i.e.* irreducible subvarieties of codimension 1. (This is well-known in algebraic geometry; see [29] for 'product formulas' with DVRs in higher dimension, analogous to the case of $k(x)$ or its finite extensions.)

Exercise 2.3.1.

(i) **Gauss norm.** Let K be a field with a DVR and associated absolute value. For a polynomial $f(x) = a_0 x^n + a_1 x^{n-1} + \ldots + a_d \in K[x]$ we put $|f(x)| := \sup |a_i|$ (Gauss norm of f). Prove that , for $f, g \in K[x]$, $|f + g| \leq \max(|f|, |g|)$ and $|fg| = |f| \cdot |g|$ (which is a 'Gauss lemma'); hence prove that this norm may be extended to $K(x)$ (with the same residue field), making $K(x)$ into a valued field with corresponding DVR.

(ii) Let K be a field with a DVR and suppose that the residue field is infinite. Recall the Gauss norm from the previous exercise and prove that $|f(x)| = \sup_{\xi \in K, |\xi|=1} |f(\xi)|$.

(iii) Let $\mathcal{O}$ be a principal domain which is a VR. Prove that $\mathcal{O}$ is a DVR. Also, let $\mathcal{O}$ be a DVR of a field L, with L containing a field K. Prove that if $\mathcal{O}$ does not contain K, $\mathcal{O} \cap K$ is a DVR of K.

(iv) In Examples 2.3.1 (i) and (ii) prove that the induced topologies (on $\mathbb{Q}$ or $k(x)$) from the various VRs are pairwise distinct. (See also below.)

(v) Verify that the following is an example of non-discrete valuation on $k(x, y)$: $|p(x, y)|_v = 2^{-a-b\sqrt{2}}$ where the polynomial $p(x, y)$ is $x^a y^b g(x, y)$ with g a polynomial coprime to xy. (By Remark 2.3.3(i) above, this absolute value induces a VR of $k(x, y)$. Prove that the corresponding maximal ideal has height 1, $i.e.$ does not strictly contain nonzero prime ideals.)

We observe that Zariski has classified valuations for finitely generated extensions K/k with transcendence degree 2 (see [8] for references); the situation is much more delicate than in the case of transcendence degree 1.

(vi) Let $\mathcal{O}$ be the subset of $K = k(x, y)$ made up of the rational functions $f(x, y)$ such that $f(0, y)$ is defined and has no pole at $y = 0$. Prove that $\mathcal{O}$ is a VR of K, but not a DVR. (Roughly speaking, it is obtained by 'composing' two places, the first on K, the second on $k(y)$.) Prove that $\mathcal{O}$ has dimension ≥ 2, $i.e.$ its maximal ideal contains properly some nonzero prime ideal. (Look at the functions f such that $f(0, y) = 0$.)

(vii) Prove that the VR of Exercise 2.2.1(i) (related to Examples 2.3.1 (iv), (v) above) is not a DVR.

(viii) Verify that $v : K^* \to \mathbb{Z}$ is a continuous surjective function (giving $\mathbb{Z}$ the discrete topology).

(ix) Let K be a field with ultrametric absolute value $|\cdot|$ and let $f(x) = x^d + a_1 x^{d-1} + \ldots + a_d \in K[x]$ have a root $\xi \in K$. Prove that $|\xi| \leq \max(1, |a_1|, \ldots, |a_d|)$.

(x) Let $|.|_v$ run through all absolute values of $\mathbb{Q}$, where the p-adic absolute value is normalised so that $|p| = p^{-1}$. Prove that for $x \in \mathbb{Q}^*$ we have the product formula $\prod_v |x|_v = 1$. Establish an analogue for $k(x)$ in place of $\mathbb{Q}$.

(xi) Let $|.|_1, |.|_2$ be absolute values on a field K. Prove that they induce the same topology on K if and only if there exists $l > 0$ such that $|x|_2 = |x|_1^l$ for every $x \in K$.

(Hint: observe first that $|x|_1 > 1$ if and only if $|x|_2 > 1$, consider powers x^n for this. Then, for any y, consider integers m, n such that $|x|_1^m > |y|_1^n$.)

2.4. Simultaneous approximations with several discrete valuation rings

In this short section we shall prove a useful result about simultaneous approximations in a field K endowed with the topologies coming from finitely many DVRs in it. This is somewhat analogue to the *Chinese Remainder Theorem*, which guarantees the solvability of simultaneous congruences modulo finitely many pairwise comaximal ideals.

We start with the simple

Proposition 2.4.1. *If $\mathcal{O} \subset \mathcal{O}'$, where $\mathcal{O}$ is a DVR of K and $\mathcal{O}'$ is a VR of K, then $\mathcal{O} = \mathcal{O}'$.*

Proof. Let $\mathcal{P}, \mathcal{P}'$ be the corresponding maximal ideals. Note that $\mathcal{P}' \cap \mathcal{O}$ is a prime of $\mathcal{O}$, hence is 0 or $\mathcal{P}$ (because $\mathcal{O}$ is a DVR). In the first case we would have $K^* \subset \mathcal{O}'^*$ and hence $K = \mathcal{O}'$, a contradiction.

Therefore $\mathcal{P}' \cap \mathcal{O} = \mathcal{P}$, and in particular $\mathcal{P} \subset \mathcal{P}'$. But now Proposition 3.2, applied with $L = K$, gives the conclusion.   □

Remark 2.4.1. In this result it suffices that $\mathcal{O}$ is a VR of K of dimension 1, *i.e.* that its maximal ideal is the only nonzero prime. If we only assume that $\mathcal{O}$ is a VR, the conclusion need not hold: see the exercises below.

Theorem 2.4.2 (Artin-Whaples. Simultaneous approximations).
Let $\mathcal{O}_1, \ldots, \mathcal{O}_n$ be pairwise distinct DVR with the same fraction field K and order functions resp. $v_1, \ldots, v_n$. Also, let $a_1, \ldots, a_n \in K$. Then, for every positive integer B there exists $x \in K$ such that $v_i(x - a_i) > B$ for $i = 1, \ldots, n$.

In particular, distinct DVRs induce distinct topologies.

Proof. We start by observing that there is no inclusion relation $\mathcal{O}_i \subset \mathcal{O}_j$ for $i \neq j$, because the $\mathcal{O}_i$'s are pairwise distinct and because of Proposition 2.4.1 above.

We go on by showing that there exists $y \in K$ having a zero at $\mathcal{O}_1$ and a pole at $\mathcal{O}_i$ for all $i = 2, \ldots, n$ (recall Def. 1.2). For $n = 2$, let $x_1 \in \mathcal{O}_1 \setminus \mathcal{O}_2$ and $x_2 \in \mathcal{O}_2 \setminus \mathcal{O}_1$. Then $y := x_1/x_2$ has a zero at $\mathcal{O}_1$ (because x_1 is regular and $1/x_2$ has a zero there) and a pole at $\mathcal{O}_2$ (because x_1 has a pole there and $1/x_2$ is either regular or has a pole), as required. In general, let $y_i, i = 2, \ldots, n$, have a zero at $\mathcal{O}_1$ and a pole at $\mathcal{O}_i$. Then it is very easy to verify that, for suitable positive integers $r_2, \ldots, r_n$, the choice $y := y_2^{r_2} + \ldots + y_n^{r_n}$ does the job. (It suffices that the pole orders at any fixed $\mathcal{O}_i$ of the $y_j^{r_j}$ are all distinct, which amounts to $(r_2, \ldots, r_n)$ be outside a finite number of hyperplanes of $\mathbb{Q}^{n-1}$.)

We now note that this yields, for any given positive B_1, a $z \in K$ having a zero of order $> B_1$ at each $\mathcal{O}_i, i = 2, \ldots, n$, and such that $z - 1$ has a zero of order $> B_1$ at $\mathcal{O}_1$; it suffices to consider $z := 1/(1 + y^r)$, for an $r > B_1$.

Finally, let z_j be an element as z, but constructed by exchanging $\mathcal{O}_1$ with $\mathcal{O}_j$. Then $x := a_1 z_1 + \ldots + a_n z_n$ is such that, for each $j = 1, \ldots, n$, $x - a_j$ has a zero of order $> B_1 - c$ at $\mathcal{O}_j$, where $c > 0$ is an upper bound for the pole-orders of the a_i at some $\mathcal{O}_j$. Taking $B_1 > B + c$ gives the sought conclusion. $\square$

Exercise 2.4.1. Construct two distinct VRs $\mathcal{O} \subset \mathcal{O}'$ of a same field K. (Hint: take any VR $\mathcal{O}$ of K of dimension ≥ 2, so its maximal $\mathcal{P}$ contains properly a nonzero prime ideal I of $\mathcal{O}$. Now, the localization $\mathcal{O}_I$ of $\mathcal{O}$ at the prime I is plainly a VR of K with maximal ideal I, containing properly $\mathcal{O}$. To find a suitable $\mathcal{O}$ use *e.g.* Exercise 2.3.1(v)).

2.5. Extension of discrete valuation rings

Definition 2.5.1. Let $K \subset L$ be fields and let $\mathcal{O} \subset \mathcal{O}'$ be DVRs in K, L respectively. Let $\mathcal{P}, \mathcal{P}'$ and v, v' be the associated primes and valuations. Assume also that $\mathcal{P} \subset \mathcal{P}'$. We say that '$\mathcal{P}'$ **lies above** $\mathcal{P}$'. We also write $\mathcal{O}'|\mathcal{O}$ or $\mathcal{P}'|\mathcal{P}$, or even $v'|v$. We call **ramification index** of $\mathcal{O}$ in $\mathcal{O}'$, or of $\mathcal{P}'$ above $\mathcal{P}$, the integer $v'(t)$, where t is a local parameter for $\mathcal{P}$. Such an integer is denoted $e(\mathcal{P}'|\mathcal{P})$ (or $e(\mathcal{O}'|\mathcal{O})$ or $e(v'|v)$). The ring $\mathcal{O}'$ (or the prime $\mathcal{P}'$) is said to be **ramified** (resp. **unramified**) above $\mathcal{O}$ if $e(\mathcal{O}'|\mathcal{O}) > 1$ (resp. $e(\mathcal{O}'|\mathcal{O}) = 1$).[6]

Finally, $\mathcal{O}$ is called a **branch point** and $\mathcal{O}'$ a **ramification point** if $\mathcal{O}'|\mathcal{O}$ is ramified.

This definition is clearly well posed: if u is another local parameter of $\mathcal{O}$ we have $t = qu$ with $q \in \mathcal{O}^* \subset \mathcal{O}'^*$ and hence $v'(t) = v'(u)$.

Also, it is immediately seen that these numbers are *multiplicative in towers*, that is

$$e(\mathcal{O}''|\mathcal{O}) = e(\mathcal{O}''|\mathcal{O}')e(\mathcal{O}'|\mathcal{O}) \qquad (2.6)$$

for DVRs $\mathcal{O} \subset \mathcal{O}' \subset \mathcal{O}''$.

We shall see several examples of these concepts. At the moment we give only a few easy ones, formulated as exercises.

---

[6] Some authors add here the condition that the residue field extension, introduced below, is separable.

Exercise 2.5.1.

(i) $K = \mathbb{Q}, L = \mathbb{Q}(\sqrt{2})$. The absolute value associated to 2 on $\mathbb{Q}$ (see Section 2.2.1, Application A) extends to L, and is ramified with index 2. All the remaining DVRs of $\mathbb{Q}$ are unramifield. (All of this follows from subsequent theorems, but it may also be easily verified directly; look at the integral closure of $\mathbb{Z}_{(2)}$ in L.)

(ii) $L = k(t), K = k(x)$ where $x = t^d$ (t an indeterminate over k, d a positive integer); now the absolute value associated to the point 0 on $k(x)$ (see Section 2.2.1, Application B) extends to the one associated to the point 0 on $k(t)$ and is ramified with index d. On the contrary the absolute value of K associated to the point 1 extends (for d prime to char(k)) to the d points of L associated to the d-th roots of unity, each extension being unramified.

(iii) Consider also an extension L/K' where now $K' = k(r(t))$, for a rational function $r \in L$. Describe the ramified points and bound their number in terms of $\deg r$.

Remark 2.5.1 (Ramified places are relevant).   The set of ramified places is an extremely important datum of an extension L/K, both in the geometric case and number-field case.

In the geometric case, say when L, K are function fields of curves (for instance over $\mathbb{C}$) the extension L/K corresponds to a topological cover of the curves corresponding to L, K (recall Ch. I), each deprived from the ramification (or branch) points. If we fix the set S of branch points in K (which will be shown to be finite in Corollary 2.6.6), it may be proved that there are only finitely many extensions L/K of given degree and unramified outside S. Moreover, the cover itself is determined up to isomorphism by the 'monodromy above S': see [27] or [35]. For instance, if $K = k(t)$ is a rational field, it will be proved that (in characteristic 0) there are no extensions of degree > 1 unramified everywhere. In the complex case, this corresponds to the fact that $\mathbb{P}_1(\mathbb{C})$ is simply connected.[7]

Similar (but usually weaker) results are known for number fields. For instance, a theorem of Minkowski says that there are no nontrivial extensions of $\mathbb{Q}$ which are unramified everywhere; hence we can say that 'Spec($\mathbb{Z}$) (the space of prime ideals) is simply connected'. (See [21], Lemma 6.5, for a generalization.) Also, it was proved by Hermite that

---

[7] In general, this may be derived as an the applications of the so-called Riemann-Hurwitz genus formula.

there are only finitely many number fields of given degree and unramified except above a prescribed finite set of primes of $\mathbb{Q}$ (see [18]).

Over number fields and function fields (over $\bar{\mathbb{Q}}$, say) it is in principle
possible to find 'effectively' the set of ramified points when we are given
concretely the defining equations, by means of factorization over complete fields.

We have now some important results about the extensions of DVRs.

Theorem 2.5.1. *Let $\mathcal{O}_K$ be a DVR of K, contained in a VR $\mathcal{O}_L$ of L,
where L/K is a finite extension. Then, if $\mathcal{P}_K$, $\mathcal{P}_L$ are the respective max-
imal ideals, we have $\mathcal{O}_K = \mathcal{O}_L \cap K$ and $\mathcal{P}_K = \mathcal{P}_L \cap K$. Hence $\mathcal{O}_L/\mathcal{P}_L$
is canonically an extension of $\mathcal{O}_K/\mathcal{P}_K$. Moreover, $\mathcal{O}_L$ is a DVR of L and
$e(\mathcal{O}_L|\mathcal{O}_K) \le [L : K]$.*

Proof. The opening argument is as in Prop. 2.4.1: $\mathcal{P}_L \cap \mathcal{O}_K$ is prime,
hence is 0 or $\mathcal{P}_K$ (because $\mathcal{O}_K$ is a DVR). In the first case, we would
have $K^* \subset \mathcal{O}_L^*$ and hence L, as an algebraic extension of K, would be
integral over $\mathcal{O}_L$; but then we would have $L = \mathcal{O}_L$ (by Proposition 2.1.3),
a contradiction.

Therefore $\mathcal{P}_L \cap \mathcal{O}_K = \mathcal{P}_K$, and in particular $\mathcal{P}_K \subset \mathcal{P}_L$. This in turn
implies $\mathcal{O}_K = \mathcal{O}_L \cap K$. In particular, we can see $\mathcal{O}_L/\mathcal{P}_L$ as an extension
of $\mathcal{O}_K/\mathcal{P}_K$.

Let now π a local parameter of $\mathcal{O}_K$ and let $t_0, \ldots, t_s \in L$ ($s \ge 1$)
with $t_0 = \pi$, $t_s = 1$ and such that $t_i/t_{i+1} \in \mathcal{P}_L$ per $i = 0, \ldots, s - 1$.
(Automatically, the t_i's lie in $\mathcal{O}_L$.)

We go on with the following

Claim: $t_1, \ldots, t_s$ *are linearly independent over K.*

In fact, suppose given a relation

$$\sum_{j=1}^{s} x_j t_j = 0, \qquad x_j \in K \text{ not all zero.} \qquad (2.7)$$

To start with, we may multiply for a suitable power of π to suppose that
the x_j, $j = 1, \ldots, s$, are all in $\mathcal{O}_K$ and are not all divisible (in $\mathcal{O}_K$)
by π. Let then $h \ge 1$ be the largest integer such that $x_h \notin \mathcal{P}_K$. Then
$x_j \in \mathcal{P}_K = \pi\mathcal{O}_K$ for $j > h$ so (2.7) entails $x_h t_h \equiv 0 \pmod{t_{h-1}\mathcal{O}_L}$.
By construction, this implies $x_h \in \mathcal{P}_L$ whence $x_h \in \mathcal{P}_L \cap K = \mathcal{P}_K$, a
contradiction which proves the Claim.

The Claim yields $s \le [L : K]$; it also follows that there exists a sequence $t_0, \ldots, t_s$ as above, in which s is maximum. Observe that for such
a sequence $t := t_{s-1}$ generates $\mathcal{P}_L$. In fact, suppose by contradiction that

$z \in \mathcal{P}_L \setminus t\mathcal{O}_L$; then $z/t \notin \mathcal{O}_L$, hence $t/z \in \mathcal{P}_L$, since $\mathcal{O}_L$ is a VR of L. But then the sequence in question would not be of maximal length, because it could be extended by inserting z between $t = t_{s-1}$ and $t_s = 1$.

Since the sequence has maximal length it also follows that $t^s \in \pi\mathcal{O}_L$ (for otherwise $\pi/t^s \in \mathcal{P}_L$ and $\pi, t^s, \dots, t, 1$ would be a longer sequence). On the other hand, the existence of this sequence implies that $\pi \in t^s\mathcal{O}_L$ (because $\pi = (\pi/t_1)(t_1/t_2)\cdots(t_{s-1}/t_s) \in \mathcal{P}_L^s$). Combining these facts we find that $\pi \in t^s\mathcal{O}_L^*$.

Let us now prove that $\cap_{n\geq 0} t^n\mathcal{O}_L = \{0\}$.[8] In fact, let y lie in this infinite intersection; as an element of $\mathcal{O}_L$, y is algebraic over K; let then $a_0 y^d + \dots + a_{d-1} y + a_d = 0$ be an equation of minimal degree, where the $a_i \in \mathcal{O}_K$ are not all zero, and not all divisible by π. The equation implies that $a_d \in \cap_{n\geq 0} t^n\mathcal{O}_L$ and, since $t^s \in \pi\mathcal{O}_L$, $a_d \in \cap_{m\geq 0} \pi^m\mathcal{O}_L$. But we have seen that $\mathcal{O}_L \cap K = \mathcal{O}_K$, whence $a_d \in \cap_{m\geq 0} \pi^m\mathcal{O}_K$ and finally $a_d = 0$ in virtue of Proposition 2.3.1. But then $y = 0$, for otherwise the degree d could be lowered. This proves that the intersection is in fact 0.

Let finally $I \subset \mathcal{O}_L$ be a nonzero ideal and let $g \in I$ be an element of minimum order at $\mathcal{P}_L$. (Note that we can define the order at $\mathcal{P}_L$ because we have shown that $\mathcal{P}_L$ is principal and because only 0 has infinite order.) Then, if $x \in I$, $x/g \in \mathcal{O}_L$, i.e. g generates I. Therefore $\mathcal{O}_L$ is principal and every ideal in it is generated by some power of t; $\mathcal{O}_L$ is then a DVR. We have seen that $\pi \in t^s\mathcal{O}_L$ so $s = e(\mathcal{O}_L|\mathcal{O}_K)$ and the inequality $s \leq [L : K]$ concludes the proof. $\qquad\square$

Definition 2.5.2. In the situation of the theorem, the degree $[\mathcal{O}_L/\mathcal{P}_L : \mathcal{O}_K/\mathcal{P}_K]$ is called *residual degree* and denoted $f(\mathcal{O}_L|\mathcal{O}_K)$ (or $f(\mathcal{P}_L|\mathcal{P}_K)$, or $f(v_L|v_K)$).

Also, we say that $\mathcal{P}$ is **totally ramified** in L when we have an equality:
$$e(\mathcal{O}_L|\mathcal{O}_K) = [L : K].$$

Remark 2.5.2. Note for instance that for extensions of function fields in one variable over an algebraically closed k the residual degree is 1, since the residue fields are equal to k, by Proposition 2.1.4.

In the situation of total ramification, the ramification index attains a maximal value by the theorem. As for the residual degree, one may also easily verify a multiplicative property
$$f(\mathcal{O}''|\mathcal{O}) = f(\mathcal{O}''|\mathcal{O}')f(\mathcal{O}'|\mathcal{O}), \qquad (2.8)$$

for DVRs $\mathcal{O} \subset \mathcal{O}' \subset \mathcal{O}''$.

---

[8] Observe that we have not yet proved that $\mathcal{O}_L$ is a DVR and we cannot therefore use Proposition 2.3.1.

We have now the following theorem, which establishes a fundamental relation among these numbers, for varying extensions of a same DVR. In particular, it implies that a totally ramified DVR admits only a single extension to L.

Theorem 2.5.2. *Let L be a finite extension of K, let $(\mathcal{O}, \mathcal{P})$ be a DVR of K and let S be the set of VRs of L containing $\mathcal{O}$. Then each $\mathcal{D} \in S$ is a DVR and $B := \cap_{\mathcal{D} \in S} \mathcal{D}$ is the integral closure of $\mathcal{O}$ in L. Also, S is a finite nonempty set and we have*

$$\sum_{\mathcal{D} \in S} e(\mathcal{D}|\mathcal{O}) f(\mathcal{D}|\mathcal{O}) = \dim_{\mathcal{O}/\mathcal{P}} \frac{B}{\mathcal{P}B} \leq [L : K]. \qquad (2.9)$$

Moreover, B is finitely generated as an $\mathcal{O}$-module if and only the last inequality is an equality. Finally, this is the case if L/K is separable.

Proof. That each $\mathcal{D} \in S$ is a DVR is a consequence of Theorem 2.5.1 and that B is the integral closure of $\mathcal{O}$ in L follows from Application E of Theorem 2.2.1 (with L in place of K and $\mathcal{O}$ in place of A). This also says that S is nonempty (which also follows directly from Theorem 2.2.1).

To go on, we fix some notation: we denote by $\mathcal{P}_\mathcal{D}$ the maximal ideal of $\mathcal{D}$ and by π (resp. $\pi_\mathcal{D}$) a local parameter of $\mathcal{O}$ (resp. of $\mathcal{D}$); also, we abbreviate $e_\mathcal{D} := e(\mathcal{D}|\mathcal{O})$, $f_\mathcal{D} := f(\mathcal{D}|\mathcal{O})$.

We continue by observing that plainly $B/\mathcal{P}B$ is indeed a vector space over $\mathcal{O}/\mathcal{P}$. Let then $b_1, \ldots, b_r \in B$ have reductions modulo $\mathcal{P}B$ which are linearly independent over $\mathcal{O}/\mathcal{P}$. We contend that they are linearly independent over K. In fact, in a relation $\sum_{i=1}^r x_i b_i = 0$, with $x_i \in K$ not all 0, we can suppose by multiplying by a suitable power of π that the x_i are in $\mathcal{O}$ and not all in $\mathcal{P}$. But then a reduction modulo $\mathcal{P}B$ gives a contradiction. Hence $\dim_{\mathcal{O}/\mathcal{P}} \frac{B}{\mathcal{P}B} \leq [L : K]$, proving the inequality in (2.9).

Suppose that we have an equality, set $d := [L : K]$ and pick $b_1, \ldots, b_d \in B$ with the property that their reductions modulo $\mathcal{P}B$ form a basis for $B/\mathcal{P}B$ over $\mathcal{O}/\mathcal{P}$; in particular, $\sum_{i=1}^d Kb_i = L$ (since we have just seen that the b_i must be linearly independent over K).

Let us prove that $B = \sum_{i=1}^d \mathcal{O}b_i$. If not, there exists $b \in B$ of the shape $b = \sum_{i=1}^d x_i b_i$, where the x_i are in K but not all in $\mathcal{O}$. Then for some positive power π^h of π we may write $\pi^h b = \sum_{i=1}^d y_i b_i$, where the $y_i = \pi^h x_i$ are in $\mathcal{O}$ but not all in $\mathcal{P}$. Then, reducing modulo $\mathcal{P}B$, we get a linear dependence over $\mathcal{O}/\mathcal{P}$ of the reductions of the b_i, a contradiction. This proves the 'if' part (since B is free over $\mathcal{O}$).

Suppose on the other hand that B is finitely generated as an $\mathcal{O}$-module. Since $\mathcal{O}$ is a DVR, whence principal, and since B is torsion free over $\mathcal{O}$

(indeed, B is a domain), B must be free over $\mathcal{O}$ by a well-known result[9] (see [18]). Let then $\omega_1, \ldots, \omega_d$ be a basis for L/K ($d = [L : K]$). Since $\pi^m \omega_j \in B$ for large enough m, necessarily B has rank d over $\mathcal{O}$, so $B \cong \mathcal{O}^d$, which immediately leads to the sought equality. (Alternatively, if $b_1, \ldots, b_r \in B$ have reductions giving a basis for $B/\mathcal{P}B$ over $\mathcal{O}/\mathcal{P}$, and if $B' = \sum \mathcal{O}b_i$, we have $B = \mathcal{P}B + B'$. Since B is finitely generated and $\mathcal{O}$ is local, by Nakayama lemma we find $B = B'$, hence necessarily $r = d$, and the sought equality follows by the independence of the b_i over K.)

Next, we prove that $\mathcal{P}B = \cap_{\mathcal{D} \in S} \mathcal{P}_{\mathcal{D}}^{e_{\mathcal{D}}}$. Note first that by definition we may write, for each $\mathcal{D} \in S$, $\pi = \pi_{\mathcal{D}}^{e_{\mathcal{D}}} \xi_{\mathcal{D}}$ for some $\xi_{\mathcal{D}} \in \mathcal{D}^*$. Let then $x \in \mathcal{P}_{\mathcal{D}}^{e_{\mathcal{D}}}$, so $x = \pi_{\mathcal{D}}^{e_{\mathcal{D}}} \eta_{\mathcal{D}}$ with $\eta_{\mathcal{D}} \in \mathcal{D}$ and by the above $x = \pi \mu_{\mathcal{D}}$ with $\mu_{\mathcal{D}} \in \mathcal{D}$. This shows that if $x \in \cap_{\mathcal{D} \in S} \mathcal{P}_{\mathcal{D}}^{e_{\mathcal{D}}}$ we have $x/\pi \in B$, proving that $\mathcal{P}B$ indeed contains the said intersection. But the opposite inclusion in fact follows from the same equalities $\pi = \pi_{\mathcal{D}}^{e_{\mathcal{D}}} \xi_{\mathcal{D}}$ for some $\xi_{\mathcal{D}} \in \mathcal{D}^*$.

Let now S' be any finite subset of S and let $B' := \cap_{\mathcal{D} \in S'} \mathcal{P}_{\mathcal{D}}^{e_{\mathcal{D}}}$. We have a natural map from $B'/\mathcal{P}B'$ to $\oplus_{\mathcal{D} \in S'} (\mathcal{D}/\mathcal{P}_{\mathcal{D}}^{e_{\mathcal{D}}})$. For $\mathcal{D} \in S'$, let $a_{\mathcal{D}} \in \mathcal{D}$ be arbitrary; by the approximation Theorem 2.4.2 we may find $x \in L$ such that $x - a_{\mathcal{D}} \in \mathcal{P}_{\mathcal{D}}^{e_{\mathcal{D}}}$ for each $\mathcal{D} \in S'$. Then x lies in B' and has reduction $a_{\mathcal{D}}$ modulo $\pi_{\mathcal{D}}^{e_{\mathcal{D}}}$, for all $\mathcal{D} \in S'$. This proves that the map is surjective.

Now, for fixed $\mathcal{D} \in S$, let $d_1, \ldots, d_{f_{\mathcal{D}}} \in \mathcal{D}$ be representatives for a basis of $\mathcal{D}/\mathcal{P}_{\mathcal{D}}$ over $\mathcal{O}/\mathcal{P}$. Then it is very easily checked that the elements $b_i \pi_{\mathcal{D}}^j, i = 1, \ldots, f_{\mathcal{D}}, j = 0, \ldots, e_{\mathcal{D}} - 1$, form a basis for $\mathcal{D}/\mathcal{P}_{\mathcal{D}}^{e_{\mathcal{D}}}$ over $\mathcal{O}/\mathcal{P}$.

Hence, by counting dimensions we deduce that $\sum_{\mathcal{D} \in S'} e_{\mathcal{D}} f_{\mathcal{D}} \leq \dim_{\mathcal{O}/\mathcal{P}} \frac{B'}{\mathcal{P}B'}$. This holds for all finite subsets S' of S. But it follows as at the beginning that $\dim_{\mathcal{O}/\mathcal{P}} \frac{B'}{\mathcal{P}B'} \leq [L : K]$. Hence we deduce that S is finite and by taking $S' = S$ we get $\sum_{\mathcal{D} \in S} e_{\mathcal{D}} f_{\mathcal{D}} \leq \dim_{\mathcal{O}/\mathcal{P}} \frac{B}{\mathcal{P}B}$. To prove that this is indeed an equality it suffices to check that the natural map from $B/\mathcal{P}B$ to $\oplus_{\mathcal{D} \in S}(\mathcal{D}/\mathcal{P}_{\mathcal{D}}^{e_{\mathcal{D}}})$ is injective. But this follows from the inclusion $\cap_{\mathcal{D} \in S} \mathcal{P}_{\mathcal{D}}^{e_{\mathcal{D}}} \subset \mathcal{P}B$, which has been proved above.

To conclude, we have to deal with the case when L/K is separable. Let $T := Tr_K^L : L \to K$ be the trace; it is a well-known easy matter to verify that $T(xy)$ is a non degenerate K-bilinear form from $L \times L$ to K, and so there exists a 'dual basis' $\omega_1^*, \ldots, \omega_d^* \in L$, so that $T(\omega_i \omega_j^*) = 1$ or 0 according as $i = j$ or not (see [18], but the existence of this is a simple fact from linear algebra). Note that $T(bb') \in \mathcal{O}$ for $b, b' \in B$, since

---

[9] This goes back essentially to Gauss.

$T(bb')$ is in K and integral over $\mathcal{O}$ (which is integrally closed). Write now an element $b \in B$ (uniquely) as $b = \sum_{i=1}^{d} x_i \omega_i$ with $x_i \in K$. We have $x_i = T(b\omega_i^*)$; since there exist an integer $N > 0$ so that $\pi^N \omega_i^*$ are all integral over $\mathcal{O}$, hence in B, we see that $\pi^N x_i \in \mathcal{O}$. This proves in particular that B is a submodule of a finitely generated module over $\mathcal{O}$, and thus is itself finitely generated, as asserted. $\qquad\qquad\qquad$ $\square$

Remark 2.5.3.

(i) That B is finitely generated as an $\mathcal{O}$-module[10] is not always true (see Exercise 2.5.2(ii) below). It is true for instance when K is *complete* w.r. to $\mathcal{P}$, or when K is a function field in one variable (over a field k), apart from the separability assumption in the theorem. (See [7].)

(ii) As a very special corollary, we obtain that $f(\mathcal{D}|\mathcal{O})$ is finite in the situation of the theorem; specializing to the case when L/k is a function field in one variable and $K = k(x)$ for an $x \in L$, x transcendental over k, we obtain another proof of Proposition 2.1.4, actually in sharpened form.

Example 2.5.1. Let K be a number field and let A_K be its ring of integers (=integral closure of $\mathbb{Z}$ in K). We can classify the VRs $(\mathcal{O}, \mathcal{P})$ of K similarly to the case $K = \mathbb{Q}$, namely by considering $P := \mathcal{P} \cap A_K$: it turns out that this is a maximal ideal of A_K and that $\mathcal{O}$ is the localization $(A_K)_P$. From Theorem 2.5.1 we find that this is always a DVR. (Note that this is subtler than in the case of $\mathbb{Q}$, because P will not be principal in general.) Conversely, each maximal ideal of A_K gives rise to a DVR by localizing. Theorem 3.1 implies that for each prime number p there is at least one such VR lying above $\mathbb{Z}_{(p)}$. The residue field will be a finite extension of $\mathbb{F}_{p^m}$ of $\mathbb{F}_p$, where m is the residual degree. Theorem 2.5.2 implies that $\sum_{\mathcal{P}|p} f(\mathcal{P}|p)e(\mathcal{P}'|p) = [K : \mathbb{Q}]$ will hold.

We illustrate this in the simple case of Exercise 2.5.1(i): Since $(\sqrt{2})^2 = 2$, the 'prime' 2 of $\mathbb{Q}$ ramifies in $\mathbb{Q}(\sqrt{2})$. After Theorem 2.5.2, the ramification index is necessarily 2 and there is only one extension of this DVR of $\mathbb{Q}$ to $\mathbb{Q}(\sqrt{2})$. Take instead the prime 7 of $\mathbb{Q}$. This extends to two DVRs of $\mathbb{Q}(\sqrt{2})$, both unramified, with local parameters $3 \pm \sqrt{2}$ and trivial residue field extension.

As to the 'infinite' absolute value of $\mathbb{Q}$ (*i.e.* the usual one), it also extends to K; the inequivalent extensions correspond to the embeddings

---

[10] Note that B is integral over $\mathcal{O}$, so it is finitely generated as an $\mathcal{O}$-module if and only if it is such as an $\mathcal{O}$-algebra.

of K in $\mathbb{C}$, up to complex conjugation; that is, for each embedding K inherits an absolute value from the one on $\mathbb{C}$. The formula of Theorem 2.5.2 remains true, interpreting as 1 the ramification indices and as 1 or 2 the residual degrees, according as the corresponding embedding of K is real or not.

Example 2.5.2 (Degree). Let now K/k be a function field in one variable; if $x \in K$ is transcendental over k, $K/k(x)$ is a finite extension. By Theorem 2.2.1 (or Theorem 2.5.2) every VR of $k(x)$ extends to (at least) a VR of K. Conversely, given a VR of K, we get one of $k(x)$ by intersection (Proposition 2.2.2). Again, Theorem 2.5.1 together with Example 2.1.2 implies that all VRs of K/k are in fact DVRs. The residue fields are finite extensions of k (by Remark (i) to Proposition 2.2.2), so the residual degrees are always 1 if k is algebraically closed.

Suppose again that k is algebraically closed. Then $\Sigma_{k(x)}$ corresponds to $\mathbb{P}_1(k)$ (Example 2.1.2). The VRs not lying above ∞ (which are finite in number: by Theorem 2.5.2 there are at most $[K : k(x)]$ of them - see below) correspond to the prime ideals of the integral closure B of $k[x]$ in K; this holds by Theorem 2.5.2. This situation is entirely analogous to the previous example (with $k[x]$ in place of $\mathbb{Z}$).

To give an explicit instance, take now the situation in Exercise 2.5.1(ii): $L = k(t), K = k(t^d)$. The point 0 of K is totally ramified in L. The point 1 instead is not ramified and admits d distinct extensions to L, all with residual degree 1.

Geometrically, the inclusion of fields $K \subset L$ corresponds to a rational map ϕ (of degree $[L : K]$) from a curve $\mathcal{C}_L$ with function field L to a curve $\mathcal{C}_K$ with function field K. (More precisely, the rational map between geometric 'models', in the sense of Chapter 1, depends on the choice of generators for K, L. For instance, the map coming from the last example is $t \mapsto t^d$.) A DVR $\mathcal{O}_K \in \Sigma_K$ corresponds to a point $P \in \mathcal{C}_K$ and the extensions of it to DVRs $\mathcal{D}$ of L correspond to the inverse image $\phi^{-1}(P) \subset \mathcal{C}_L$. When the ground field k is algebraically closed, the residual degrees are 1, and Theorem 2.5.2 substantially says that '*the number of inverse images, counted with multiplicity, is equal to the degree of the map ϕ.*' The multiplicity is measured by the ramification index, and is 1 (as we shall prove) for all but finitely many points of $\mathcal{C}_K$.

Remark 2.5.4. As alluded in Chapter 1, in the case $k = \mathbb{C}$, it may be shown that the DVR of $K/\mathbb{C}$ correspond to the points of a 'compact Riemann surface' $\mathcal{R}$ (that is, a compact connected complex variety of dimension 1); we plan to touch this situation in a later volume, but see

[7] or [16]. A local parameter at $\mathcal{O}$ gives a biholomorphic map from a neighborhood of $\mathcal{O}$ in $\mathcal{R}$ to an open disk around 0 in $\mathbb{C}$.

We can sum up a (small) part of these conclusions in the following:

Corollary 2.5.3. *The VRs of number fields and function fields in one variable K/k are DVRs.*

Remark 2.5.5. We shall see later another proof of this important corollary. As remarked above, we shall also see that in these cases there are only a finite number of DVRs of K ramified in a given finite separable extension L.

We further note that Corollary 2.5.3 and Theorem 2.5.2 immediately give an alternative proof of Proposition 2.2.2, because we find that the residual degree is finite, whence the residual extension is algebraic and hence trivial.

Another important and natural consequence is the following

Corollary 2.5.4. *If K/k is a function field in one variable, each $x \in K^*$ has only finitely many zeros and poles. More precisely, if x is transcendental over k it has at least one zero (and one pole) and at most $[K : k(x)]$ zeros (or poles).*

Similarly, if K is a number field any $x \in K^*$ has only finitely many zeros and poles.

Proof. On replacing x with $1/x$, we see that it it sufficient to prove only the assertion for zeros. Let $(\mathcal{O}, \mathcal{P})$ be a DVR of K which is a zero of x; this means that $x \in \mathcal{P}$. Then x must be transcendental over k, for otherwise $1/x$ would lie in $\mathcal{O}$, because, being algebraic over k, it would be integral over $\mathcal{O}$.

Consider then the DVR $\mathcal{O}_0$ of $k(x)$ corresponding to the point $0 \in k$; it is the unique DVR of $k(x)$ where x has a zero; plainly, $\mathcal{O} \supset \mathcal{O}_0$. Conversely, if $x \in \mathcal{O}$ and $x(\mathcal{O}) \neq 0$, then $1/x \in \mathcal{O} \setminus \mathcal{O}_0$. Therefore the DVR of K above $\mathcal{O}_0$ are precisely the zeros of x, and now Theorem 2.5.2 yields immediately the conclusion (also about existence of one zero in case x is transcendental over k, since S is then nonempty).

The argument is somewhat similar for number fields: if $(\mathcal{O}, \mathcal{P})$ is a zero of x, then $x \in \mathcal{P}$; let $a_0 x^d + \ldots + a_{d-1} x + a_d = 0$ be a nontrivial equation for x over $\mathbb{Z}$, where we may assume that $a_d \neq 0$. Then $a_d \in \mathcal{P} \cap \mathbb{Z} = p\mathbb{Z}$, say, where p is a suitable prime number. This means that the integer a_d is divisible by p; this is true only for finitely many primes, and for each such prime p there are at most $[K : \mathbb{Q}]$ possibilities for $\mathcal{P}$, by Theorem 2.5.2, which concludes the argument. $\qquad \square$

Note that for number fields we cannot assert the existence of a zero; in fact, the so called *units* in a number field are characterized as the elements with no zeros or poles. They form a group of finite rank, predicted by a famous theorem by Dirichlet (see [18, 21]). Naturally, the result for number fields amounts to saying that an algebraic number has a numerator (or denominator) which is divisible by at most finitely many prime ideals. Related to these remarks is the following important

Definition 2.5.3. *S-integers and S-units.* Let S be a subset of Σ_K. We define the ring $\mathcal{O}_S = \mathcal{O}_{K,S}$ of S-integers in K as the set of elements of K whose only possible poles (in Σ_K) are in S; in other words $\mathcal{O}_S = \{x \in K : |x|_v \leq 1, \ \forall v \in \Sigma_K \setminus S\}$.

We also define the group of S-units as $\mathcal{O}_S^*$; it is the set of elements of K whose only poles and zeros lie in S.

Note that indeed $\mathcal{O}_S$ is a ring; by Corollary 2.5.4, each element of K lies in some $\mathcal{O}_S$ for a suitable *finite* S.

In the case of number fields (resp. function fields/k) these rings are finitely generated (resp. algebras/k). In the case of $\mathbb{Q}$ and $k(x)$ this is rather easy to see; for finite extensions, in the separable case this may be proved as in Theorem 2.5.2 above, using the trace map.

As an instance, if $K = \mathbb{Q}$, the ring $\mathcal{O}_S$ is the set of rational numbers whose denominator is made up only of primes from S, while the S-units are those rationals whose numerator and denominator are of this shape. When S is empty, $\mathcal{O}_\emptyset^*$ is the above mentioned *group of units*. For $K = \mathbb{Q}$ it consists of ± 1, but it is nontrivial for other number fields; for instance for $K = \mathbb{Q}(\sqrt{2})$ this group is $\{\pm(1 + \sqrt{2})^m \ : \ m \in \mathbb{Z}\}$. In general, a celebrated result (proved by Dirichlet for $S = \emptyset$) asserts that $\mathcal{O}_S^*$ is the product of a finite group of roots of unity by a free abelian group of rank $\#S + a - 1$, where a is the number of archimedean absolute values of K (we have $a = r + c$ where r, c are resp. the number of real and complex embeddings of K, up to complex conjugation).

As another instance, if $K = k(x)$, $S = \{\infty\}$, the ring $\mathcal{O}_S$ is the polynomial ring $k[x]$. In general, in the function field case the group $\mathcal{O}_{K,S}^*$ has finite rank modulo k^* but there is not a general formula as simple as Dirichlet's for this rank. The question is related to the rank of the group generated by the linear equivalence classes of the points in S.

We conclude this section with the simple but important

Proposition 2.5.5. *Let $K \subset L$ be fields with DVRs $\mathcal{O}_K \subset \mathcal{O}_L$ and primes $\mathcal{P}_K \subset \mathcal{P}_L$. Then the topology on L corresponding to $\mathcal{P}_L$ induces on K the topology corresponding to $\mathcal{P}_K$.*

Proof. It suffices to note that if $x \in K$, $v_L(x) = e v_K(x)$, where e is the ramification index. Hence the ratio of the order functions is bounded above and below by positive constants, and the assertion follows.   □

Exercise 2.5.2.

(i) Develop another proof of part of Theorem 2.5.2 as follows: for a fixed finite subset S' of S and for each $\mathcal{D} \in S'$, select elements $a_{i\mathcal{D}} \in \mathcal{D}$, $i = 1, \ldots, f_\mathcal{D}$, whose reductions modulo $\mathcal{P}_\mathcal{D}$ form a basis over $\mathcal{O}/\mathcal{P}$. Then, with the aid of Theorem 2.4.2, pick the local parameter $\pi_\mathcal{D}$ so that it has 'large' order at all $\mathcal{D}' \in S'$, $\mathcal{D}' \neq \mathcal{D}$. Now prove that the $a_{i\mathcal{D}} \pi_\mathcal{D}^j$, $i = 1, \ldots, f_\mathcal{D}$, $j = 0, 1, \ldots, e_\mathcal{D} - 1$, are linearly independent over K.[11]

(ii) Provide details in the following example of a field K with a DVR $\mathcal{O}$ and a quadratic extension L/K such that the integral closure B of $\mathcal{O}$ in L is not finitely generated over $\mathcal{O}$. Let k be a field of characteristic $p > 0$, t be an indeterminate over k, $\varphi(t) := \sum_{m=0}^{\infty} a_m t^m \in k[[t]]$ a series which is transcendental over $k(t)$.[12] Let $K := k(t, \varphi^p)$, $L := k(t, \varphi)$. Prove that $[L : K] = p$. Embed $K \subset L$ in $k((t))$ and let $\mathcal{O} := K \cap k[[t]]$, so $\mathcal{O}$ is a DVR of K. Define, for any positive integer N, $\psi_N := (\varphi - \sum_{m=0}^{N-1} a_m t^m)/t^N \in L$ and prove that ψ_N is integral over $\mathcal{O}$. (Consider ψ_N^p.) Now conclude by proving that if B were finitely generated as an $\mathcal{O}$-module, then the coefficient c_1 of φ in a (possible) representation $b = c_0 + c_1\varphi$, $c_0, c_1 \in K$, of an element $b \in B$ would have necessarily a pole of bounded order at $\mathcal{O}$.
Prove directly that $\dim_{\mathcal{O}/\mathcal{P}}(B/\mathcal{P}B) = 1$ now.

(iii) For a given prime number p, consider the DVR $\mathcal{O}$ of $\mathbb{Q}$ associated to p (*i.e.* $\mathcal{O} = \mathbb{Z}_{(p)}$). Prove that $\mathcal{O}$ may be extended to some VR $\mathcal{O}'$ of $\overline{\mathbb{Q}}$ and that no such $\mathcal{O}'$ can be a DVR. Also, prove that there exist some infinite degree algebraic extensions L of $\mathbb{Q}$ such that each VR of L extending $\mathcal{O}$ is a DVR. Provide analogous conclusions for $k(x)$ in place of $\mathbb{Q}$.

(iv) Assumptions and notations being as in Theorem 2.5.2, assume moreover that L/K is a finite Galois extension, with group G. Show that G acts transitively on S.
(Hint: Let $\mathcal{D}, \mathcal{D}' \in S$. If $\mathcal{D}' \neq \sigma(\mathcal{D})$, for all $\sigma \in G$, find $x \in L$, $x \equiv 0$

---

[11] This argument need not to introduce the ring B.

[12] It exists, as can be seen either by cardinality considerations or by taking *e.g.* a suitable series with large gaps.

(mod $\mathcal{D}'$), $x \equiv 1$ (mod $\sigma(\mathcal{D})$) for all $\sigma \in G$. Then the norm $N_K^L(x)$ lies in $\mathcal{D}'$ hence in $\mathcal{O}$, hence in $\mathcal{D}$, a contradiction.)

(v) Context being as in (iv), deduce that all the ramification indices and residual degrees of the primes $\mathcal{D}|\mathcal{O}$ are equal (to integers e, f resp.), and in particular the degree $[L : K]$ equals efr, where r is the number of DVRs lying above $\mathcal{O}$.

(vi) **(Eisenstein polynomials)** Let K be a field with DVR $(\mathcal{O}, \mathcal{P})$ and let $f(X) = X^n + a_1 X^{n-1} + \ldots + a_n \in \mathcal{O}[X]$, where $a_i \in \mathcal{P}$ for all i but $a_n \notin \mathcal{P}^2$. Prove that f is irreducible and that a root defines a totally ramified extension of K.

2.6. Completions of discrete valuation rings

We have seen in Section 2.3 that a field K with a DVR $(\mathcal{O}, \mathcal{P})$ becomes also equipped with associated order function, absolute value and distance. Therefore, like for any metric space, we can consider the **completion** $\hat{K}$ of K with respect to this distance. As a matter of notation, this will be sometimes denoted by K_v, if v denotes the relevant DVR or valuation.

A field K complete with respect to a DVR is also called a **local field**.

Recall that $\hat{K}$ may be defined as the *quotient space of the set of* Cauchy sequences *in K modulo the equivalence relation which identifies two sequences $\{a_n\}$ and $\{b_n\}$ such that $d(a_n, b_n) \to 0$.* Since K is a field, we can add sequences and so we can also say that $\hat{K}$ is the quotient space of the group of Cauchy sequences modulo the subgroup consisting of sequences converging to 0. (Recall that a sequence $\{a_n\}_{n \in \mathbb{N}}$ is said to be Cauchy if $d(a_m, a_n) \to 0$ as $\min(m, n) \to \infty$.)[13]

It is easily verified that if $\{a_n\}$, $\{b_n\}$ are Cauchy sequences in K, the real sequence $\{d(a_n, b_n)\}$ converges; the limit defines a quasi-distance between the Cauchy sequences and induces a distance on $\hat{K}$. The original space K embeds isometrically as a dense subset of $\hat{K}$ by associating to $x \in K$ the sequence with constant value x. Also, $\hat{K}$ is found to be complete.

It is also easily verified that the space $\hat{K}$ becomes a topological field under componentwise operations of representative sequences.

We finally recall that any closed subset of $\hat{K}$ is automatically complete.

We refer to any book on general topology for the easy proofs of these assertions.

---

[13] The group of Cauchy sequence is a ring under the natural operations and the subgroup of those converging to 0 may be easily verified to be a maximal ideal.

Remark 2.6.1. An equivalent construction for $\hat{K}$ occurs with projective limits (see *e.g.* [2] or [26]); more precisely, we can consider the 'projective limit of the rings $\mathcal{O}/\mathcal{P}^n$', for $n \to \infty$. This consists of the sequences $\{b_n\}_{n\in\mathbb{N}}$, $b_n \in \mathcal{O}/\mathcal{P}^n$, verifying the compatibility conditions $b_{n+1} \equiv b_n$ (mod $\mathcal{P}^n$) for all $n \in \mathbb{N}$.

Proposition 2.6.1. *Let $\{x_n\}$ be a sequence in K, converging to $x \in \hat{K}$. Then, if $x = 0$, we must have $v(x_n) \to +\infty$. If $x \neq 0$, $v(x_n)$ is eventually constant, in $\mathbb{Z}$. The closure $\hat{\mathcal{O}}$ of $\mathcal{O}$ in $\hat{K}$ is a closed DVR of $\hat{K}$. Its maximal ideal is the closure $\hat{\mathcal{P}}$ of $\mathcal{P}$ and is unramified above $\mathcal{P}$. Finally, $\mathcal{O}/\mathcal{P}$ embeds isomorphically onto $\hat{\mathcal{O}}/\hat{\mathcal{P}}$.*

Proof. Since $\{x_n\}$ converges in $\hat{K}$, it is a Cauchy sequence, *i.e.* $v(x_m - x_n) \to +\infty$ as $\min(m, n) \to \infty$. Suppose first that $v(x_n) \to +\infty$; then $x_n \to 0$ by definition, so $x = 0$. Conversely, if $x = 0$ then plainly $v(x_n) \to +\infty$. In the remaining cases, there are an $l \in \mathbb{Z}$ and an infinite subsequence x_{m_j} ($m_1 < m_2 < \ldots$) such that $v(x_{m_j}) \leq l$ for all $j \in \mathbb{N}$. For large enough j, n we have $v(x_n - x_{m_j}) > l$, so by (4.2) applied with $a = x_n - x_{m_j}$, $b = x_{m_j}$, we have $v(x_n) = v(x_{m_j})$. This shows that $v(x_n)$ is eventually constant (in $\mathbb{Z}$), as asserted.

This also says that we can define an order function $v(x)$ as the limit of $v(x_n)$; plainly this is independent of the sequence $\{x_n\}$ in K converging to x. (In particular, the order function and distance function on $\hat{K}$ take the same values as on K.) It is also clear that for $x \in \hat{\mathcal{O}}$ (resp. $x \in \hat{\mathcal{P}}$) we have $v(x) \geq 0$ (resp. $v(x) \geq 1$). Letting t be a local parameter of $\mathcal{O}$, we then have $x/t \in \hat{\mathcal{O}}$ for all $x \in \hat{\mathcal{P}}$, proving that $\hat{\mathcal{P}} = t\hat{\mathcal{O}}$. Also, one verifies that an ideal I of $\hat{\mathcal{O}}$ is generated by any element in it of minimum order. It follows that all nonzero ideals are powers of $\hat{\mathcal{P}}$, so $\hat{\mathcal{O}}$ is a DVR. The fact that a local parameter of $\mathcal{O}$ is a local parameter of $\hat{\mathcal{O}}$ proves completely everything but the last assertion. But this is obvious because $\mathcal{O}$ is by definition dense in $\hat{\mathcal{O}}$. $\square$

Remark 2.6.2.

(i) Note that a series $\sum_{n=0}^{\infty} x_n$, $x_n \in \hat{K}$, converges if and only if $x_n \to 0$; this holds because the absolute value is ultrametric (so for $s > r$ we have $|\sum_{n=r}^{s} x_n| \leq \sup_{r \leq n \leq s} |x_n|$). We also have $|\sum_{n=0}^{\infty} x_n| \leq \sup |x_n|$, with equality if the supremum is attained precisely once. (Use Proposition 2.3.4.)

(ii) We also find that $\hat{K}$ consists of the series $\sum_{n=0}^{\infty} x_n$, where $x_n \in K$, $x_n \to 0$. (In fact, if $x \in \hat{K}$ is $\lim y_n$, for a Cauchy sequence $\{y_n\}$

in K, it suffices to set $x_0 = y_0$ and $x_n := y_n - y_{n-1}$ for $n \geq 1$.)
Actually, more is true, as in the following proposition.

Proposition 2.6.2. *Let for $n \in \mathbb{N}$, R_n be a system of representatives in $\hat{O}$ for $\hat{O}/\hat{P}$ and let $\hat{t} \in \hat{K}$ be a local parameter. Then each $x \in \hat{O}$ may be written uniquely as a series $\sum_{n=0}^{\infty} r_n \hat{t}^n$, where $r_n \in R_n$.*

Proof. Uniqueness: Let $\sum_{n=0}^{\infty} r_n \hat{t}^n = \sum_{n=0}^{\infty} r'_n \hat{t}^n$, where $r_n, r'_n \in R_n$, and let m be an index such that $r_m \neq r'_m$. If m is minimum, then we see that $r_m \equiv r'_m \pmod{\hat{P}}$, a contradiction.

Existence: Let $x \in \hat{O}$. We show by induction on $d \geq 0$ that there are $r_0, r_1 \ldots, r_n \in R_n$, such that $v(x - \sum_{n=0}^{d-1} r_n \hat{t}^n) \geq d$. This will suffice because then the corresponding infinite series will plainly converge to x.

For $d = 0$ the claim holds because $x \in \hat{O}$. Assuming this to be true for d, we can write $x - \sum_{n=0}^{d-1} r_n \hat{t}^n = \hat{t}^d y$, where $y \in \hat{O}$. It then suffices to choose $r_d \in R_d$ such that $y \equiv r_d \pmod{\hat{P}}$ to get the property for $d + 1$ in place of d. $\qquad\square$

Example 2.6.1. Let p be a prime number and $\mathbb{Z}_{(p)}$ be the DVR of $\mathbb{Q}$ associated to it. The corresponding completion is classically denoted $\mathbb{Q}_p$ and is called the field of *(rational) p-adic numbers*. The corresponding complete DVR is denoted $\mathbb{Z}_p$ and is called the ring of *p-adic integers*.

The fact that the residue field $\mathbb{F}_p$ is finite implies that $\mathbb{Z}_p$ is compact (and $\mathbb{Q}_p$ is locally compact).

Using the last Remark(ii) it is easily seen that each element of $\mathbb{Q}_p$ is representable as a series $\sum_{n=r}^{\infty} a_n p^n$, for some $r \in \mathbb{Z}$, where the a_n are suitably taken from a(ny) fixed system of representatives in $\mathbb{Z}$ for $\mathbb{Z}/(p)$.[14]

Similarly, if K is a number field and P is a prime ideal of its ring of integers $\mathcal{O}_K$, we have seen that the localization $(\mathcal{O}_K)_P$ is a DVR (Ex. 5.2) and we can consider the corresponding completion $\hat{K}$ of of K, denoted sometimes by K_P, or K_v, if v is the corresponding absolute value. If P lies above the prime number p, K_P is a finite extension of $\mathbb{Q}_p$. We shall see that $[K_P : \mathbb{Q}_p]$ is the product of the ramification index times the residual degree. The finite extensions of $\mathbb{Q}_p$ are usually called *p*-**adic fields**.

---

[14] A standard system of representatives consists of the integers $0, 1, \ldots, p - 1$; but we may also go outside $\mathbb{Z}$ and choose the union of 0 with the set of $(p - 1)$-th roots of unity, which indeed lie in $\mathbb{Z}_p$ (as shall be proved). This last system has the property that the nonzero elements form a multiplicative group; they are called *Teichmüller* representatives. See [24] for generalisations of this canonical choice.

Exercise 2.6.1.

(i) Using that $\mathbb{Z}_p$ is compact prove that, for a polynomial $f \in \mathbb{Z}[X]$, the equation $f(x) = 0$ has a solution in $\mathbb{Z}_p$ if and only if for every integer n the congruence $f(x) \equiv 0 \pmod{p^n}$ has a solution in $\mathbb{Z}$. (Similarly for several variables.)

(ii) Prove that if p, q are distinct primes the fields $\mathbb{Q}_p$ and $\mathbb{Q}_q$ are not isomorphic.
(Hint: if $q > 2$ consider *e.g.* the equation $x^2 = p(p + q)$.) Prove also that they are not isometric. [15]

(iii) Prove that $6 \sum_{m=0}^{\infty} 7^m = -1$ in $\mathbb{Q}_7$ and also that $2 \sum_{m=0}^{\infty} \binom{1/2}{m} (-5/4)^m$ converges to a square root of -1 in $\mathbb{Q}_5$.

(iv) Prove that $2 + \frac{2^2}{2} + \frac{2^3}{3} + \ldots$ converges to 0 in $\mathbb{Q}_2$.
(Hint: look at $\log(1 - x)$ for $x = 2$ and try to develop some properties of the series for the exponential and logarithmic function over $\mathbb{Q}_p$. See for instance [5] or [10].)

(v) Prove that for $p \neq 2$, $\mathbb{Q}_p$ has precisely three quadratic extensions: $\mathbb{Q}_p(\sqrt{a})$, $\mathbb{Q}_p(\sqrt{p})$ and $\mathbb{Q}_p(\sqrt{ap})$, where $a \in \mathbb{Z}$ is any quadratic non-residue modulo p. What happens for $p = 2$? (Use *e.g.* (i); a quick method is provided by Proposition 2.6.3 below.)
Prove that $\mathbb{Q}_p$ contains infinitely many fields $\mathbb{Q}(\sqrt{m})$, $m \in \mathbb{Z}$. [16]

(vi) Prove that the binomial coefficients $\binom{x}{m}$, $m \in \mathbb{N}$, take values in $\mathbb{Z}_p$ for $x \in \mathbb{Z}_p$.

(vii) Let M be an $r \times r$ matrix over $\mathbb{Z}_p$, such that $M \equiv I \pmod{p}$ and $M \equiv I \pmod 4$ if $p = 2$. Prove that if M has finite order, then $M = I$.
(Hint: Write $M = I + p^h N$ where N is a matrix over $\mathbb{Z}_p$ and $h \in \mathbb{Z}$ is maximal.)
Deduce that a subgroup of $GL_r(\mathbb{Z}_p)$ consisting of elements of finite order must be a finite group. (Other proofs of this result may be given by character theory.)

Example 2.6.2. Let now k be algebraically closed, fix $a \in \mathbb{P}_1(k)$ and let $\mathcal{O}$ be the VR of $k(x)$ associated to it. Using Proposition 2.6.2, the

---

[15] On the other hand, it is known, and not difficult to see, that all the $\mathbb{Z}_p$ are homeomorphic to the Cantor set, as topological spaces (so any two among the $\mathbb{Q}_p$, or $\mathbb{Z}_p$, are homeomorphic). For instance, here is a homeomorphism $f : \mathbb{Z}_3 \to \mathbb{Z}_2$: let $x = a_0 + 3a_1 + 3^2 a_2 + \ldots$ be a 3-adic integer, where $a_i \in \{0, 1, 2\}$. Put $f(x) = b_0 + 2b_1 + \ldots$, where $b_0 b_1 \ldots$ comes from $a_0 a_1 \ldots$ by transforming an $a_i = 0$ into 0, an $a_i = 1$ into 10 and an $a_i = 2$ into 11.

[16] One can show, using 'Krasner's lemma', that there are only finitely many extensions of $\mathbb{Q}_p$ of given degree; see the exericises below or [18]. Similar results hold for other complete fields.

completion $\hat{K}$ is easily seen to be isomorphic to the field $k((x - a))$ (resp. $k((1/x))$ if $a \in k$ (resp. if $a = \infty$). The associated valuation ring is $k[[x - a]]$ (resp. $k[[1/x]]$).

More generally, if K is a function field in one variable over $k(x)$, the completion $\hat{K}$ at a DVR $\mathcal{O}$ of K is isomorphic to $k((t))$ if t is (any) local parameter of $\mathcal{O}$.

Exercise 2.6.2.

(i) **Puiseux series.** Let k be algebraically closed of characteristic 0. Prove that there is only one extension of $k((x))$ of degree d, namely $k((u))$ where $u^d = x$.

(Hint: Let L be the given extension of $K = k((x))$; choose a DVR of L extending the canonical one of K and embed L in the corresponding completion $\hat{L}$.[17] Choosing $t \in L$ to be a local parameter, we have $\hat{L} = k((t))$, by Proposition 2.6.2. We have also $x = t^e s(t)$ with $s \in k[[t]]$ and $s(0) \neq 0$, where e is the ramification index. Writing $s(t) = s(0) + t s_1(t)$ with $s_1 \in k[[t]]$ and extracting an e-th root of s with the binomial theorem (or using Proposition 2.6.3 below) leads to $x = (t \cdot r(t))^e$ with $r \in k[[t]]$, $r(0) \neq 0$. Thus $k((t))$ contains an e-th root u of x with a local parameter $u = t \cdot r(t)$. Thus $k((t)) = k((u))$ (by Proposition 2.6.2 again, or inverting $u = t \cdot r(t)$). Since L has degree d over $k((x))$ and is contained in $k((u))$, we must have $d \leq e$; but by Theorem 2.5.1, $e \leq d$, so actually $d = e$ and $L = k((u))$, as wanted.) This extension is Galois with cyclic group.

(ii) Show that the assumption $\text{char}(k) = 0$ cannot be omitted, even if we restrict to separable extensions. Let k be an algebraic closure of $\mathbb{F}_p$. Show that the equation $Y^p - Y + (1/x) = 0$ defines a separable extension of degree p of $k((x))$, but has no solutions in $k((x^{1/d}))$, no matter the integer d. (The assumption $\text{char}(k) = 0$ may be modified to 'char(k) not dividing d'. In that case, char(k) does not divide e, because $e|d$ by Theorem 2.6.8 below, and the above proof works.)

Remark 2.6.3. The result at part (i) in particular implies that the algebraic closure of $k(x)$ may be embedded in the union of the fields $k((x^{1/d}))$, for varying integer $d \geq 1$. Classically, this is referred to as 'Puiseux Theorem'; it may also be proved without appeal to completions, constructing the power-series solutions by means of 'Newton polygons' (see [1, 10, 36]); this approach has the advantage of being 'effective', in the sense that it allows to compute the relevant ramification index e (provided

---

[17] By Theorem 2.6.8 below we have $\hat{L} = L$ but we do not need this now.

we work with "effective" equations). Note that when x is a local parameter at 0, but we want the expansion 'at infinity', we may choose $1/x$ as a local parameter, and then the expansion takes the form $c_a x^{a/d} + c_{a-1} x^{(a-1)/d} + \ldots$, that is, in decreasing powers of x; this is often the formulation of Puiseux result.

The power-series solutions $y = y(x^{1/d})$ of algebraic equations $f(x,y) = 0$ may be shown to have a positive radius of convergence when $f \in k[X, Y]$ and k is a *valued field* (see Appendix B below or see [1], [10]). When k is a number field their coefficients have interesting arithmetical properties: for instance their denominators are divisible by only finitely many primes (Eisenstein Theorem); see Appendix B for a discussion and a proof.

When $k = \mathbb{C}$ the result is a formal analogue of the fact that holomorphic maps between Riemann surfaces may be viewed locally as the map $t \mapsto t^d$.

In complete valued fields we have a most important criterion for analyzing roots of algebraic equations; it is called '**Hensel's lemma**' and substantially in its simplest version it amounts to looking at the reduction modulo $\mathcal{P}$ of the equation and then lifting the roots modulo arbitrarily high powers of $\mathcal{P}$. We prove here a version of the lemma for factors of arbitrary degree (see *e.g.* [10] for a more general one).

We shall denote with a tilde the reduction modulo $\mathcal{P}$.

Proposition 2.6.3. *Let $\mathcal{O}$ be a DVR of the field K, supposed to be complete w.r. to the associated valuation. Let $f \in \mathcal{O}[x]$ and suppose that $\tilde{f} \equiv g_0 h_0 \pmod{\mathcal{P}}$, where $g_0, h_0 \in \mathcal{O}[x]$, g_0 is monic and $\tilde{g}_0, \tilde{h}_0$ are coprime in $(\mathcal{O}/\mathcal{P})[x]$. Then there exist $g, h \in \mathcal{O}[x]$ with g monic, $f = gh$ and $\tilde{g} = \tilde{g}_0, \tilde{h} = \tilde{h}_0$.*

Proof. Let π be a local parameter for $\mathcal{O}$. We are going to construct inductively a sequence of polynomials $g_n, h_n \in \mathcal{O}[x]$ such that g_n is monic, $\tilde{g}_n = \tilde{g}_0, \tilde{h}_n = \tilde{h}_0$, $f \equiv g_n h_n \pmod{\pi^{n+1}}$ and $\deg g_n = \deg g_0$, $\deg h_n \leq \deg f - \deg g_0$.

For $n = 0$ we take the polynomials in the statement. Suppose g_n, h_n given and put $f - g_n h_n =: \pi^{n+1} \phi$ with $\phi \in \mathcal{O}[x]$.

We shall define $g_{n+1} = g_n + \pi^{n+1} \gamma$, $h_{n+1} = h_n + \pi^{n+1} \delta$, for suitable $\gamma, \delta \in \mathcal{O}[x]$ with $\deg \gamma < \deg g$.

Since $\tilde{g}_0, \tilde{h}_0$ are coprime in $(\mathcal{O}/\mathcal{P})[x]$, we may pick γ, δ such that $\gamma h_0 + \delta g_0 \equiv \phi \pmod{\pi}$; by replacing γ with its remainder upon division by g_0 it is also possible to assume that $\deg \gamma \leq \deg g_0 - 1$. Hence $\deg \delta \leq \max(\deg \phi - \deg \tilde{g}_0, \deg h_0 - 1) \leq \deg f - \deg g_0$, by induction.

It is immediately verified that this definition settles the inductive step. Also, the construction shows that, since $\mathcal{O}$ is complete, g_n, h_n converge to $g, h \in \mathcal{O}[x]$ satisfying the conclusion. $\qquad\square$

A significant special case occurs when $\deg g_0 = 1$. This corresponds to $\tilde{f}$ having a simple root; see examples 2.6.3 below for applications. We can refine further this case by means of an analogue of the well-known 'Newton's method' in elementary analysis, for the real roots of equations (whether algebraic or not).

Proposition 2.6.4. *Let $\mathcal{O}$ be a DVR of the field K, supposed to be complete w.r. to the associated valuation. Let $f \in \mathcal{O}[X]$ and suppose that $u \in \mathcal{O}$ is such that $\lambda := |f(u)|/|f'(u)|^2 < 1$. Then the recurrence sequence $u_0 = u, u_{n+1} = u_n - (f(u_n)/f'(u_n))$ is well defined and converges to $\rho \in \mathcal{O}$ such that $f(\rho) = 0$ and $|\rho - u| \leq \lambda |f'(u)| < 1$.*

Further, ρ is the unique root of f in K at distance $\leq \lambda |f'(u)|$ from u.

Proof. Put $\delta_n := -f(u_n)/f'(u_n)$. We prove by induction on $n \geq 0$ that δ_n is defined (*i.e.*, $f'(u_n) \neq 0$) and that:

(i$_n$) $|u_n - u| \leq \lambda |f'(u)|$.
(ii$_n$) $|f'(u_n)| = |f'(u)|$.
(iii$_n$) $|\delta_n| \leq |f'(u_n)|\lambda^{2^n}$.

For $n = 0$ these facts are contained in the assumptions. Suppose them to be true up to n and let us prove them with $n + 1$ in place of n.

To start with, we have $u_{n+1} := u_n + \delta_n$ by definition, so $|u_{n+1} - u| \leq \max(|u_{n+1} - u_n|, |u_n - u|) \leq \max(|\delta_n|, \lambda |f'(u)|) = \lambda |f'(u)|$, by (i$_n$), (ii$_n$) and (iii$_n$). This proves (i$_{n+1}$).

Next, since f has coefficients in $\mathcal{O}$, $|f'(u_{n+1}) - f'(u_n)| \leq |u_{n+1} - u_n|$; in turn, this is $\leq |\delta_n| \leq \lambda |f'(u_n)|$, by (iii$_n$). Hence (using the last assertion of Proposition 2.3.4) and (ii$_n$) we get (ii$_{n+1}$).

Note that this implies that δ_{n+1} is well-defined.

Finally, by Taylor expansion we find (again noticing that $f \in \mathcal{O}[X]$) $|f(u_{n+1})| = |f(u_{n+1}) - f(u_n) - \delta_n f'(u_n)| \leq |\delta_n^2|$. Using (iii$_n$) and (ii$_{n+1}$) (which has been just verified using only the n-th step) this yields in turn $|f(u_{n+1})| \leq |f'(u_n)^2 \lambda^{2^{n+1}}| = |f'(u_{n+1})^2 \lambda^{2^{n+1}}|$. On dividing by $|f'(u_{n+1})|$ we then obtain (iii$_{n+1}$), completing the induction.

Now, property (iii$_n$) implies that $\delta_n \to 0$ for $n \to \infty$, whence $\{u_n\}$ is a Cauchy sequence in K, and thus converges to $\rho \in K$. By (i$_n$) we have $|\rho - u| \leq \limsup |u_n - u| \leq \lambda |f'(u)| < 1$. Also, by (iii$_n$) again, $|f(u_n)| = |\delta_n| \cdot |f'(u_n)| \leq |\delta_n| \to 0$, so $f(\rho) = 0$. This completes the proof of the first part.

For the last part, let $\xi \in K$ be a root of f with $|\xi - u| \leq \lambda|f'(u)|$ and set $\xi = u_n + \eta_n$. We prove by induction that $|\eta_n| \leq |f'(u)|\lambda^{2^n}$. For $n = 0$ this is true by assumption, so suppose it true up to n. We have $0 = f(\xi) = f(u_n) + \eta_n f'(u_n) + O(\eta_n^2)$, by Taylor expansion. Dividing by $f'(u_n)$ we get $|\eta_n - \delta_n| \leq |\eta_n|^2/|f'(u)|$ (by (ii$_n$)). But $\eta_{n+1} = \eta_n - \delta_n$, so the conclusion follows from the inductive assumption.

We have proved in particular that $|\xi - u_n| \to 0$ and since $u_n \to \rho$ we get the full statement. $\qquad\qquad\square$

The propositions apply *e.g.* when $|f(u)| < 1$ and $|f'(u)| = 1$; this amounts to saying that *the reduction $\bar{u} \in \mathcal{O}/\mathcal{P}$ of u modulo $\mathcal{P}$ is a simple root of the reduction $\bar{f} \in (\mathcal{O}/\mathcal{P})[X]$ of f*. We shall now see a few simple instances of this.

Example 2.6.3.

(i) Let p be a prime > 2 and let $a \in \mathbb{Z}$ be coprime to p and a *quadratic residue of p*, *i.e.*, the congruence $x^2 \equiv a \pmod{p}$ has a solution $x = x_0 \in \mathbb{Z}$. Then a is a square in $\mathbb{Q}_p$: this follows from Prop. 7.3, with $f(X) = X^2 - a$, $u = x_0$; in fact, $f'(x_0) = 2x_0$ is invertible modulo p because $p \nmid 2a$. For $p = 2$ this does not work, but Proposition 2.6.4 still implies that a is square in $\mathbb{Q}_2$ if $a \equiv 1 \pmod{8}$: now we may set $u = 1$ and observe that $f(1)$ is divisible by 8 whence $|f(1)|_2 < |f'(1)|_2^2$.

Another interesting example comes from the equation $x^{p-1} = 1$ over $\mathbb{F}_p$. This has $\mathbb{F}_p^*$ as a set of solutions, all simple. Then we can lift them to $\mathbb{Q}_p$, which is thus found to contain all the $(p-1)$-th roots of unity.

(ii) Similar applications are to equations in several variables. For instance, take $p > 2$ and a quadratic equation $\sum_{i=1}^m a_i X_i^2 = 0$, $a_i \in \mathbb{Z}_p$ not all divisible by p; *there exists a nontrivial solution $X_i = x_i \in \mathbb{Z}_p$ if there is a solution to the corresponding congruence modulo p, with not all the entries divisible by p*.

(iii) As to the completions of function fields in one variable, we obtain power series fields $k((x))$, where x is a local parameter of the relevant DVR. Take *e.g.* the case of the point $x = 0$ of $k(x)$; then $\widehat{k(x)} = k((x))$; now, we may see a polynomial $f \in \mathcal{O}[X]$ as an element $f(x, Y) \in k[[x]][Y]$ in two variables. Reduction modulo $\mathcal{P}$ amounts to the substitution $x = 0$ and so the proposition yields for instance the following corollary: *If $f(0, a) = 0$ but $f_Y(0, a) \neq 0$, $(a \in k)$, there exists a unique formal solution $Y = Y(x) \in k[[x]]$ to $f(x, Y(x)) = 0$ with $Y(0) = a$*. This is a formal version of Dini's implicit function theorem. (See Appendix B below and compare with Exercise 2.6.2, where, in char 0, the existence of solutions in $k((x^{1/d})$ is guaranteed, independently of the present assumptions.)

Also, we see that the distinct roots of $f(0, Y)$ give rise to factors of $f(x, Y)$ over $k((x))$.

A further useful application of Prop. 2.6.3 is the following

Criterion for absence of ramification

We have:

Theorem 2.6.5. *Let $\mathcal{O}_K$ be a DVR of K, contained in a DVR $\mathcal{O}_L$ of L, where L/K is a finite extension, and let $\mathcal{P}_K, \mathcal{P}_L$ be the respective maximal ideals. Suppose that $L = K(y)$ where $f(y) = 0$, for a monic $f \in \mathcal{O}_K[X]$ such that $f'(y) \notin \mathcal{P}_L$. Then $\mathcal{P}_L$ is unramified over $\mathcal{P}_K$.*

Proof. Let $\hat{L}$ be the completion of L w.r. to $\mathcal{P}_L$ and consider the closure of K in $\hat{L}$. Since $\mathcal{P}_L$ lies above $\mathcal{P}_K$ (by Theorem 2.5.1) this closure coincides with $\hat{K}$, the completion w.r. to $\mathcal{P}_K$ (by Proposition 2.5.5).

In the rest of the proof, let us denote with a tilde the reduction modulo $\hat{\mathcal{P}}_L$. We have $\tilde{f}(\tilde{y}) = 0$; let $\tilde{g}_0 \in (\mathcal{O}_L/\mathcal{P}_L)[X]$ be the minimal polynomial of $\tilde{y}$ over $\mathcal{O}_K/\mathcal{P}_K$, the reduction of a monic polynomial $g_0 \in \mathcal{O}_K[X]$, of degree $d = \deg \tilde{g} \leq f(\mathcal{O}_L|\mathcal{O}_K)$. (Recall that the residue field extensions are unchanged under completion.)

We may write $\tilde{f} = \tilde{g}_0\tilde{h}_0$ for some $h \in \mathcal{O}_K[X]$. If $\tilde{g}, \tilde{h}$ were not coprime, then $\tilde{g}_0$ would divide $\tilde{h}_0$, whence $\tilde{y}$ would be a double root of $\tilde{f}$, which is not the case by assumption. Therefore by Proposition 7.3 we have a factorization $f = gh$ as in that statement, where $g, h \in \hat{\mathcal{O}}_K[X]$. Since $\tilde{h}(\tilde{y}) \neq 0$, we have $g(y) = 0$.

Therefore $H := \hat{K}(y)$ is a subfield of $\hat{L}$, of degree $\leq d$ over $\hat{K}$; its intersection with $\mathcal{O}_{\hat{L}}$ is a DVR $(\mathcal{O}_H, \mathcal{P}_H)$ of H lying above $\mathcal{O}_{\hat{K}}$ (Proposition 2.2.2 and Theorem 2.5.1); since $\mathcal{O}_H/\mathcal{P}_H$ contains $\tilde{y}$, its residual degree over $\hat{K}$ is at least d. By Theorem 2.5.2, it is unramified and has degree exactly d over $\hat{K}$. Hence its closure $\hat{H}$ in $\hat{L}$ is also unramified over $\hat{K}$. (It turns out, see Proposition 2.6.7 below, that H is automatically closed, but we do not need this now.)

Since $y \in \hat{H}$, $\hat{H}$ contains L and therefore the closure of L, which is just $\hat{L}$. Hence $\hat{H} = \hat{L}$ and in particular $e(\mathcal{O}_L|\mathcal{O}_K) = 1$. □

Remark 2.6.4.

(i) When $f(\mathcal{O}_L|\mathcal{O}_K) = 1$ (*i.e.* the residue fields of $\mathcal{O}_K$ and $\mathcal{O}_L$ are the same), or even when $\tilde{y} \in \mathcal{O}_K/\mathcal{P}_K$, the proof simplifies, because we do not need to construct H, and $\hat{K}$ plays the role of $\hat{H}$.

(ii) Observe that the condition $f'(y) \notin \mathcal{P}_L$ of the theorem is by no means necessary for absence of ramification: Take for instance the extension L of $K = k(x)$ determined by a root y of $f(x, Y) := Y^3 + Y^2 - x^2$, and take the points (=DVR) $x = 0$ of K and $y = 0$ of L. The root $Y = 0$ of $f(0, Y)$ is not simple, but nevertheless $\mathcal{O}_L$ is unramified over $\mathcal{O}_K$ (as can be seen from the parametrization $x = t^2 - 1$, $y = t^3 - t$ which gives $k(x, y) = k(t)$, so $L/K = k(t)/k(t^2-1)$. (With this description, L/K is defined by a different, quadratic, equation, $i.e.$ $t^2 - 1 - x = 0$. The points above $x = 0$ are $t = \pm 1$.)

2.6.1. On valuation rings and geometric points again

Let $L = k(x, y)$, where x is transcendental over k (alg. closed) and $f(x, y) = 0$, for an irreducible $f \in k[X, Y]$. We have remarked above that if $\mathcal{O}$ is a DVR of L containing x, y (this holds for 'almost all' the DVRs of L, by Corollary 2.5.4, then $(x(\mathcal{O}), y(\mathcal{O})) \in k^2$ is a point on the affine curve determined by f. And in Application B to Theorem 2.2.1 we have seen that each 'geometric' point $(a, b) \in k^2$ with $f(a, b) = 0$ 'comes' in this way from $at\ least$ one DVR of L.

Now, from Theorem 2.6.5 and its proof we may deduce that this DVR is $unique$ if $f_Y(a, b) \neq 0$. In fact, Let $K = k(x)$ and let $\mathcal{O}_L$ be a DVR of L with $a = x(\mathcal{O}_L), b = y(\mathcal{O}_L)$. Then $\mathcal{O}_L$ lies above the DVR $x = a$ of K; also, by Examples 2.6.3 (or Propositions 2.6.3, 2.6.4) there is a unique formal series $y(x) \in k[[x-a]]$ with $f(x, y(x)) = 0$ and $y(a) = b$. (This is implicit in the proof of Theorem 2.6.5; note that $H = \hat{K}$ now.) The proof of Theorem 2.6.5 also shows that $x - a$ is a local parameter for $\mathcal{O}_L$. Note that each such $\mathcal{O}_L$ gives rise to a series $y(x)$ as above, as y lies in the completion which is $k[[x - a]]$, and we have seen that this series is unique. Then the $\mathcal{O}_L$-valuation of an element $r(x, y) \in L$ is determined by the order at $x = a$ of $r(x, y(x))$. This proves uniqueness.

In practice, the theorem says that for testing non-ramification in an extension defined by an equation, it is sufficient to check that the reduced polynomial modulo $\mathcal{P}_K$ has only simple roots. This yields $e.g.$ the following

Corollary 2.6.6. *If K/k is a function field in one variable over the algebraically closed field k, there are only finitely many DVRs which are ramified in a given finite separable extension L/K.*

Proof. It is well known that one may find $x \in K$ such that $K/k(x)$ is separable (see [18]). By the multiplicative property (2.6) of ramification indices, we then reduce to the case $K = k(x)$. Now, write $L = K(y)$ where y is a root of the monic irreducible equation $f(Y) = 0$, for an $f \in K[Y] = k(x)[Y]$ (we may do this because L/K is separable). If $\mathcal{O}$ is a DVR of K, not a pole of x, then $\mathcal{O}$ corresponds to $x_0 \in k$. Then, by the theorem, $\mathcal{O}$ may be ramified in L only if the class $y_0 \in k$ of y modulo $\mathcal{P}_L$ is a double root of $f(x_0, Y)$; in this case the discriminant of f w.r. to Y (which is nonzero because f is separable, and so has no multiple factors) vanishes at x_0. Since by Corollary 2.5.4 there are only finitely many relevant poles or zeros, altogether we find only finitely many ramified $\mathcal{O}$'s, concluding the argument. □

The result holds actually without the assumption that k is algebraically closed, and also for number fields, with an entirely similar proof.

Exercise 2.6.3.

(i) For a given integer d, find all the DVRs of $\mathbb{Q}$ ramified in $L = \mathbb{Q}(\sqrt{d})$.
(ii) Find all the DVRs of $\mathbb{C}(x)$ ramified in $L = \mathbb{C}(x, y)$, where $y^3 + xy + x^2 = 0$.

We have now an important general fact; to state it, we recall first that if K is a field equipped with a nontrivial (that is, not constantly $= 1$ on K^*) absolute value $|\cdot|$[18] and if V is a vector space over K, a *norm* on V is a function $||\cdot|| : V \to \mathbb{R}^+$, such that (i) $||v|| = 0$ if and only if $v = 0$, (ii) $||cv|| = |c|\,||v||$ for $c \in K$, $v \in V$, (iii) $||v + w|| \le ||v|| + ||w||$ for $v, w \in V$.

We say that two such norms $||\cdot||_1, ||\cdot||_2$ on V are equivalent if they induce the same topology on V. It is easy to see that this amounts to inequalities

$$c||v||_1 \le ||v||_2 \le c'||v||_1,$$

for suitable positive constants c, c'.[19]

Exercise 2.6.4. Prove the last statement. (One will have to use that $|\cdot|$ is not trivial, *i.e.* not constantly 1 on K^* to ensure that, for $v \ne 0$, the set $\{||cv|| : c \in K\}$ intersects some *fixed* interval $[\delta, 1]$, with $0 < \delta < 1$. In detail, suppose for

---

[18] In this discussion it is not important to assume that $|\cdot|$ is ultrametric and discrete-valued; that is, we only assume that K is non-discrete, that $|x| = 0$ if and only if $x = 0$, that $|xy| = |x||y|$ and the usual triangle inequality.

[19] Note that this condition for equivalence is somewhat different compared to that of absolute values on a field. Indeed, in the first place here the absolute value on the field is already given, and clearly affects the norm; also, the triangle inequality would not necessarily hold on taking powers.

instance that there exists a sequence (v_n) in V with $||v_n||_2 > n||v_n||_1$. We may assume that $c_3 \geq ||v_n||_2 \geq c_4 > 0$ for suitable constants c_3, c_4 (on changing if needed v_n with $t_n v_n$, for a suitable t_n in K^*). Then $v_n \rightarrow_1 0$, but $v_n \nrightarrow_2 0$.))

Proposition 2.6.7. *Let K be a field, complete under an absolute value $|\cdot|$ and let V be a K-vector space of finite dimension. Then all norms on V are equivalent. As a consequence, V is complete w.r to any norm.*

Proof. We note at once that the last assertion follows indeed from the rest, since K^n is complete w.r. to the natural topology.

Let now $\omega_1, \ldots \omega_d$ be a basis of V/K; we prove that any norm $||\cdot||$ is equivalent to the 'sup-norm $|\cdot|$' defined by $|\sum_{i=1}^{d} x_i \omega_i| := \sup |x_i|$, for $x_1, \ldots, x_d \in K$.

Plainly we have $||\sum_{i=1}^{d} x_i \omega_i|| \leq (\sum_{i=1}^{d} ||\omega_i||) \sup |x_i|$, proving one half of the desired inequality.

To prove the second half we argue by induction on d, the conclusion being plainly true when $d = 1$. We then assume it true up to $d - 1$.

We argue by contradiction. If there does not exist a positive constant c such that $|v| \leq c||v||$ for all $v \in V$, we can find an infinite sequence $v_1, v_2, \ldots \in V \setminus \{0\}$ such that $||v_n||/|v_n| \rightarrow 0$ for $n \rightarrow \infty$. Write $v_n = \sum_{i=1}^{d} x_{in} \omega_i$, for $x_{in} \in K$. Going to an infinite subsequence we may assume that $|x_{dn}| = |v_n|$ and dividing v_n by x_{dn} we may further assume that $x_{dn} = 1$ and $|v_n| = 1$ for each n. Hence we may write $v_n = \omega_d + v'_n$ where v'_n lies in the space spanned by $\omega_1, \ldots, \omega_{d-1}$ and $|v'_n| \leq 1$. Since $||v_n|| \rightarrow 0$, we have $||v'_n - v'_m|| \rightarrow 0$ for $\min(m, n) \rightarrow \infty$. Since we are assuming the result for $d - 1$ in place of d, this yields $|x_{in} - x_{im}| \rightarrow 0$ for large m, n, i.e. $\{x_{in}\}_{n \in \mathbb{N}}$ is a Cauchy sequence, for each $i = 1, \ldots, d - 1$. Since K is complete, there exist $x_1, \ldots, x_{d-1} \in K$ with $x_{in} \rightarrow x_i$ for $i = 1, \ldots, d - 1$. Then, setting $v := \omega_d + \sum_{i=1}^{d-1} x_i \omega_i$ we have $|v_n - v| \rightarrow 0$, so $||v_n - v|| \rightarrow 0$; but $||v_n|| \rightarrow 0$, so $||v|| = 0$, implying $v = 0$, a contradiction which finally proves what we want. $\square$

Remark 2.6.5. The assumption that K is complete cannot be dropped. Simple counterexamples arise *e.g.* from the cases $K = \mathbb{Q}$ and $K = k(t)$, as shown in the nest exercises.

Exercise 2.6.5.

(i) Let k be a field, t be transcendental over k and set $x = t^2$, $K = k(x)$, $V = k(t)$, so that V is a K-vector space of dimension 2. Let $\mathcal{O}$ be the DVR of K associated to the point $x = 1$ (see Example 2.3.1(ii)) and let $|\cdot|$ be the associated absolute value. Define then the norms $||\cdot||_{\pm}$ on V as the absolute values associated to the points $t = \pm 1$ (both of which extend $|\cdot|$). Prove that these norms are not equivalent.

(ii) Similarly to (i), analyze the situation by replacing K with $\mathbb{Q}$, V with $\mathbb{Q}(\sqrt{-1})$, and letting $| \cdot |$ be the 5-adic norm on $\mathbb{Q}$ and $|| \cdot ||_{\pm}$ on V be associated to the primes $2 \pm \sqrt{-1}$, lying above 5.

(iii) Let $K = \mathbb{Q}$, $V = \mathbb{Q}(\sqrt{2})$ and let $| \cdot |$ be the usual absolute value restricted to V. Let also $| \cdot |_{-}$ be defined on V by $|a + b\sqrt{2}|_{-} = |a - b\sqrt{2}|$, for $a, b \in \mathbb{Q}$. Prove that $| \cdot |, | \cdot |_{-}$ give inequivalent extensions of the usual absolute value on $\mathbb{Q}$.

Theorem 2.6.8. *Let K be a field with a DVR $\mathcal{O}$, complete w.r. to it, and let L be a finite extension of K. Then:*

(i) *The valuation of K may be uniquely extended to L. There is only one DVR $\mathcal{D}$ of L lying above $\mathcal{O}$. It is the integral closure of $\mathcal{O}$ in L and L is complete w.r. to $\mathcal{D}$.*

(ii) *$\mathcal{D}$ is finitely generated as an $\mathcal{O}$ module, and we have*

$$e(\mathcal{D}|\mathcal{O})f(\mathcal{D}|\mathcal{O}) = [L : K].$$

Proof. Let $| \cdot |_1, | \cdot |_2$ be two valuations of L extending the given one on K. By Proposition 2.6.7, they induce equivalent norms on L, viewed as a finite dimensional vector space over K. Therefore $|x|_1 \leq c|x|_2$ for a fixed positive number c and each $x \in L$. Therefore $|x|_1^n \leq c|x|_2^n$ for each integer n, which implies that $|x|_1 \leq |x|_2$. By symmetry it follows that the valuations are in fact equal.

In particular, since distinct DVRs have associated valuations which induce distinct topologies (Theorem 2.4.2) there is only one DVR $\mathcal{D}$ in L above $\mathcal{O}$; by Theorem 2.5.2 it is the integral closure of $\mathcal{O}$ in L. Again by Proposition 2.6.7, L is automatically complete w.r. to the valuation associated to $\mathcal{D}$.

This proves part (i).

By Theorem 2.5.2 we have $e(\mathcal{D}|\mathcal{O})f(\mathcal{D}|\mathcal{O}) \leq [L : K]$ and in particular the residual degree f is finite. Let $t_1, \ldots, t_f \in \mathcal{D}$ have reductions modulo $\mathcal{P}_\mathcal{D}$ which are a basis for $\mathcal{D}/\mathcal{P}_\mathcal{D}$ over $\mathcal{O}/\mathcal{P}$ and let $\pi_\mathcal{D}$ (resp. π) be a local parameter of $\mathcal{D}$ (resp. of $\mathcal{O}$). We prove that $\mathcal{D}$ coincides with the $\mathcal{O}$-module M generated by the products $t_i \pi_\mathcal{D}^j$, $i = 1, \ldots, f$, $j = 0, \ldots, e - 1$.

First we note that every element z of $\mathcal{D}$ can be written as $x_1 t_1 + \ldots + x_f t_f + \pi_\mathcal{D} z_1$ for some $x_i \in \mathcal{O}$ and some $z_1 \in \mathcal{D}$. Applying this to z_1 in place of z and continuing e times we see that $z = m_0 + \pi z'$, where $m_0 \in M$ and $z' \in \mathcal{D}$. Since $\mathcal{D}$ is complete we have $z = \sum_{r=0}^{\infty} m_r \pi^r$ for $m_r \in M$ and this yields easily the assertion, since $\mathcal{O}$ is complete.

Finally, by Theorem 2.5.2, we have an equality in (2.9), so $ef = [L : K]$. $\square$

Remark 2.6.6. In the function field case, one may view geometrically the uniqueness of a point $\mathcal{D}$ above $\mathcal{O}$ in the complete case by observing that 'locally' different points above $\mathcal{O}$ belong to different connected components. More precisely, suppose that L/K is a finite extension of function fields over k; this corresponds to a rational map between the curves corresponding to L, K, taking $\mathcal{D}$ to $\mathcal{O}$. Now, say that $e(\mathcal{D}|\mathcal{O}) = 1$. Then the curves corresponding to K, L are 'analitically isomorphic in neighborhoods of $\mathcal{O}$, $\mathcal{D}$ resp.'. This means that the local rings of 'formal functions' (*i.e.* the formal power series) around $\mathcal{O}$, $\mathcal{D}$ are the same and that we have a (local) analytic map between the curves in the opposite direction, sending $\mathcal{O}$ to $\mathcal{D}$. This in turn implies that different points above $\mathcal{O}$ belong to different *irreducible components* (and different connected components) from the local-analytic point of view. (Over $\mathbb{C}$ it is also true that there are holomorphically equivalent neighborhoods of $\mathcal{O}$, $\mathcal{D}$.) Naturally, if we do not take completions there does not exist a local 'rational' isomorphism if $[L : K] > 1$.

As an instance (referring also to Example 1.2.3), take the elliptic field $L = k(x, y)$, $y^2 = x^3 + 1$ and $K = k(x)$, corresponding to a non-rational curve E (of genus 1) and the projective line $\mathbb{P}_1$ with a map $x : E \to \mathbb{P}_1$. The point $x = 0$ of K lifts to $(0, \pm 1)$. As we have seen, the field $k(E) = L$ is not unirational, so there is no non-constant *rational* map from $\mathbb{P}_1$ to E. However there exists an (invertible) *analytic* map from a neighborhood of 0 on the line onto a neighborhood of $(0, 1)$ on E, inverse to the map x. This is expressed by the power series $x \mapsto (x, \sqrt{1 + x^3})$ expanded by the binomial theorem. (Note that the image of the neighbourhood by the map $x \mapsto (x, -\sqrt{1 + x^3})$ yields another connected component.)

Corollary 2.6.9. *Let K be a field with a DVR $\mathcal{O}$, complete w.r. to it. Then the induced valuation may be uniquely extended to an algebraic closure $\bar{K}$ of K. If $\sigma \in \mathrm{Aut}(L/K)$, where L/K is an algebraic extension, then $|\sigma(x)| = |x|$ for all $x \in \bar{K}$. Also, if L/K is finite of degree d, we have $|x| = |N_K^L(x)|^{1/d}$.*

Proof. By Theorem 2.6.8, the valuation of K may be extended uniquely to every extension L/K of finite degree. Then, one easily verifies that it may be extended uniquely to any algebraic extension and in particular to the whole of $\bar{K}$. For $\sigma \in \mathrm{Aut}(L/K)$ the position $||x|| := |\sigma(x)|$ defines a valuation on L extending the one on K; hence we deduce $|\sigma(x)| = |x|$ by uniqueness. Finally, the norm $N_K^L(x)$ is a product of d conjugates of x, which can be extended as automorphisms of a normal closure of L/K, and the required formula follows. $\qquad\square$

Theorem 2.6.10. *Let K be a field with a DVR $\mathcal{O}$, let $\hat{K}$ be its completion and let Ω be an algebraic closure of $\hat{K}$. Further, let L be a finite extension of K. Then the DVRs $\mathcal{D}$ of L lying over $\mathcal{O}$ correspond bijectively to the embeddings of L in Ω over K up to isomorphism over $\hat{K}$, where we associate to an embedding the DVR $\mathcal{D}$ corresponding to the induced topology on L.*

Also, if $\hat{L}_{\mathcal{D}}$ is the completion of L w.r. to $\mathcal{D}$ and we embed $\hat{K}$ and L in $\hat{L}_{\mathcal{D}}$ in the natural way, then $\hat{K} \cdot L = \hat{L}_{\mathcal{D}}$. We have $\sum_{\mathcal{D}|\mathcal{O}}[\hat{L}_{\mathcal{D}} : \hat{K}] \le [L : K]$, with equality under the conditions of Thm 6.2 above, in particular if L/K is separable.

Proof. If $\mathcal{D}|\mathcal{O}$ we can embed $\hat{L}_{\mathcal{D}}$, and thus L, in Ω over K, as a finite extension of $\hat{K}$. Plainly, embeddings $\varphi_1, \varphi_2 : L \to \Omega$ that differ by an isomorphism over $\hat{K}$ induce on L the same topology, because of Corollary 2.6.9.

Conversely, suppose that the embeddings induce on L the same topology and define $L_i := \varphi_i(L) \subset \Omega$. Let $z_1 \in \hat{L}_1$ and write $z_i = \lim \varphi_1(x_n)$ where $x_n \in L$; then x_n is a Cauchy sequence in L for the topology coming from φ_1 and so $\varphi_2(x_n)$ converges, say to $z_2 \in \hat{L}_2$. We leave it to the reader to verify that the map $z_1 \mapsto z_2$ is an isomorphism from $\hat{L}_1$ to $\hat{L}_2$ over $\hat{K}$, transforming φ_1 to φ_2.

For the second assertion, observe that $\hat{K} \cdot L$ is complete by Theorem 2.6.8, whence it coincides with $\hat{L}_{\mathcal{D}}$. The last assertion also follows from Theorem 2.5.2, taking into account the equality $e(\mathcal{D}|\mathcal{O}) f(\mathcal{D}|\mathcal{O}) = [\hat{L}_{\mathcal{D}} : \hat{K}]$, coming from Theorem 2.6.8 (applied with $\hat{L}_{\mathcal{D}}, \hat{K}$ in place of L, K). $\square$

Corollary 2.6.11. *Assumptions being as in Theorem 2.6.10, suppose that L/K is separable. Let $T := Tr_K^L$ be the trace map and let $T_{\mathcal{D}} = Tr_{\hat{K}}^{\hat{L}_{\mathcal{D}}}$. Then $T = \sum_{\mathcal{D}|\mathcal{O}} T_{\mathcal{D}}$ holds on L. Similarly, $N = \prod_{\mathcal{D}|\mathcal{O}} N_{\mathcal{D}}$ holds for the norm maps.*

Proof. Let $z \in L$. Then $T(z)$ is the sum of the conjugates of z over K, namely the sum $\sum_f f(z_i)$ where f runs through the distinct embeddings of L/K in Ω. We can partition these embeddings into classes, putting in a same class two embeddings f, g such that $f = h \circ g$ for an $h \in Gal(\Omega/\hat{K})$. By Theorem 2.6.10, the classes correspond to the DVR $\mathcal{D}$ lying above $\mathcal{O}$. Also, the embeddings in the class corresponding to $\mathcal{D}$ induce embeddings of $\hat{L}_{\mathcal{D}}$ in Ω over $\hat{K}$. Thus the corresponding sum is just $T_{\mathcal{D}}(z)$. This proves the first part of the statement, the second one for the norms being completely similar. $\square$

Remark 2.6.7.

(i) We can see all of this more explicitly in the case of a primitive separable extension L/K given by $L = K(y)$, where $f(y) = 0$ for a monic irreducible $f \in K[Y]$ (hence without multiple factors).
We factor f over the completion $\hat{K}$:

$$f(Y) = \prod_{i=1}^{r} f_i(Y), \qquad f_i \in \hat{K}[Y],$$

with monic, irreducible (over $\hat{K}$) and pairwise distinct $f_1, \ldots, f_r$.
Let $d := \deg f$, $d_i := \deg f_i$ and let y_i be a root of f_i. Then $L_i := \hat{K}(y_i)$ is an extension of $\hat{K}$ of degree d_i, and is complete by Theorem 2.6.8. To each f_i we can then associate a DVR $\mathcal{D}_i$ of L, lying above $\mathcal{O}$. If e_i, f_i are the ramification index and residual degree at $\mathcal{D}_i$ we have $e_i f_i = d_i$ and $d = \sum d_i = \sum e_i f_i$, as predicted by Theorem 2.5.2.

(ii) Another equivalent formulation of these facts is by means of tensor products (see [6], Ch. II] or [24]). Namely, we can consider $L \otimes_K \hat{K}$; in the separable case this is found to be isomorphic (as a K algebra, algebraically and topologically) to the direct sum of the fields L_i/K introduced just above:

$$L \otimes_K \hat{K} = \oplus_{i=1}^{r} L_i.$$

Example 2.6.4. The factorization of a polynomial over the completion may be often 'found' in practice by a method boiling down to Hensel's Lemma, namely by working with congruences modulo a sufficiently high power $\mathcal{P}^h$ of $\mathcal{P}$, factoring over the residue rings $\mathcal{O}/\mathcal{P}^h$ and then lifting to $\mathcal{O}$.

In the case when L/K is an extension of number fields there is an effective procedure for finding such factorizations. This is not completely obvious, because even if the residue fields are finite we need some 'a priori estimate for a power of p which is large enough to ensure that the factorization in the residue ring lifts. (See Exercise 2.6.7(viii) below.) For instance, take the simple case of $f(X) = X^2 - 2$ over $\mathbb{Q}_2$. The reduction modulo 2 is reducible, but we cannot lift it. In fact, we have no factorization modulo 4 and a fortiori in $\mathbb{Z}_2$.

In the case of function fields L/k in one variable, the situation is similar. We may choose a generation $L = k(x, y)$, with $F(x, y) = 0$ a monic irreducible equation for y over $k(x)$. Then we may expand the solutions in Puiseux series (see Exercise 2.6.2), which gives the ramification indices. (Note also that when k is algebraically closed, the ramification indices are given by the degrees of the factors, because the residual degrees must be 1.) A method relying on Newton polygons ([1, 10]) allows to determine the indices in finitely many steps (provided of course we can make 'effective' calculations over the ground field). See also Appendix B for further discussion.

Example 2.6.5 (Gonality and Lüroth's theorem again). All of this also leads to a somewhat different proof of Lüroth Theorem 1.1.2. Before this, let us give a definition.

Definition 2.6.1. Let K be a function field in one variable over the algebraically closed field k. We define the **gonality** of K/k, denoted here $\gamma = \gamma(K) = \gamma(K/k)$, as the minimal degree of a nonconstant element $x \in K$, namely the minimal $[K : k(x)]$ for $x \in K \setminus k$.

For instance, a rational field is of the shape $k(x)$ and the function x has thus degree 1; hence $\gamma(k(x)) = 1$. Conversely, if $\gamma(K) = 1$ there exists a function x of degree 1, which just means that $K = k(x)$ and K is rational. The elliptic field of Example 1.2.3, defined as $k(x, y)$, where $y^2 = f(x)$ for a cubic polynomial f without multiple factors, is not rational (as shown in Theorem 1.2.1), hence $\gamma(K) > 1$. The function x has degree 2, hence $\gamma(K) = 2$. (In general, function fields of curves with $\gamma = 2$ are called *hyperelliptic*.)

Exercise 2.6.6. Let $K = k(x, y)$ where $x^d + y^d = 1$, be the function field of the Fermat curve of degree d, where d is an integer ≥ 2 coprime to char(k). (See Exercise 1.4.1.) Prove that $\gamma(K) = d - 1$.
(Hint: Note that every function on K may be expressed essentially uniquely as $f = (c_0(x) + c_1(x)y + \ldots + c_{d-1}(x)y^{d-1})/q(x)$, where $c_0, \ldots, c_{d-1}, q$ are coprime polynomials. On counting poles, prove that, if a is the maximum index such that $c_a \neq 0$ and $b = \deg q$, we have $\deg f \geq \max(a, b)(d-a)$; this implies $\gamma(K) \geq d - 1$. For the opposite inequality, consider, e.g., functions $x - cy$, with $c^d + 1 = 0$.)

The gonality is an invariant of the function field, certainly not as relevant as the *genus*[20], which however we plan to touch only in a subsequent volume of notes. Nevertheless, the gonality too may yield sometimes interesting informations.

---

[20] It may be shown that if g is the genus, then $\gamma \leq (g + 1)/2$.

We prove that the gonality is non-decreasing under field extensions (a property which is shared by the genus).

Theorem 2.6.12. *Let* $K \subset L$ *be a extension of function fields in one variable over* k. *Then* $\gamma(K) \leq \gamma(L)$.

Proof (sketch). Let $x \in L$ be a nonconstant rational function, and write a minimal equation $f(x) = 0$ over K, where $f(X) = X^d + a_1 X^{d-1} + \ldots + a_d, a_i \in K$. Some of the coefficients a_i shall not be in k, and let us select a_m as one of them. We are going to prove that

$$[K : k(a_m)] \leq [L : k(x)],$$

which will clearly suffice, for we may assume that x is a function in L of minimal degree, *i.e.* such that $[L : k(x)] = \gamma(L)$.

Recall that, by Theorem 2.5.2, the degree may be interpreted as the number of poles counted with multiplicity. Let then P be a pole of a_m, with multiplicity μ, and let $z \in K$ be a local parameter at P.

We may factor f over the completion $\hat{K}$ at P, obtaining $f = f_1 \cdots f_r$, as above, with $f_i \in \hat{K}[X]$ irreducible over $\hat{K}$, of degree $d_i = e_i$, *i.e.* equal to the ramification index, since the residual degrees are 1 now. Let $v = v_P$ be the normalized valuation (w.r. to K) associated to P.

We have $v(a_m) = -\mu$. If Q_i denotes the point of L corresponding to f_i, and if x has a pole at Q_i of order μ_i, we have $v(f_i(0)) = -\mu_i$, because all the roots of f_i have the same valuation (in view of Corollary 2.6.9); for the same reason all the coefficients of f_i have a valuation which is at least $-\mu_i$. Hence the Gauss norm (in the additive sense) of f_i is $-\mu_i$. By Exercise 2.3.1, we have $v(f) = -\sum \mu_i$. However $v(a_m) = -\mu \geq v(f)$, proving that $\sum \mu_i \geq \mu$. Summing now over all the poles of a_m we immediately obtain the claim. $\square$

We now deduce Lüroth Theorem; note that for simplicity we are assuming here k algebraically closed (unlike the proof given above in 1.1.2), but such restriction could be eliminated also in the present approach.

Corollary 2.6.13 (Lüroth Theorem). *If* $k \subsetneq K \subset k(x)$, *then* $K = k(z)$ *for a suitable* $z \in K$.

Proof. It suffices to remark that $\gamma(k(x)) = 1$, hence by the previous theorem we have $\gamma(K) \leq 1$ so $\gamma(K) = 1$ and then if $z \in K$ has degree 1 we have $K = k(z)$. $\square$

Theorem 2.6.10 implies a further important result:

Corollary 2.6.14. *Let K be a field with a DVR $\mathcal{O}$ and let L_1/K, L_2/K, L/K be finite extensions such that $L = L_1 L_2$. Suppose that $\mathcal{O}$ is unramified in both L_1, L_2 (i.e. each place above $\mathcal{O}$ is unramified) and that all the residue field extensions above $\mathcal{O}$ are separable. Then $\mathcal{O}$ is unramified in L.*

Proof. Let $\mathcal{D}$ be a place of L above $\mathcal{O}$; we may embed L, hence L_1 and L_2, in the completion $\hat{L}$ at $\mathcal{D}$. In this embedding we have, by Theorem 7.3, $\hat{L} = \hat{K} \cdot L = \hat{K} \cdot L_1 L_2 = \hat{L}_1 \cdot \hat{L}_2$, where $\hat{K}$, $\hat{L}_1$, $\hat{L}_2$ are the closures of K, L_1, L_2 in $\hat{L}$, and are therefore complete. Thus it suffices to prove the conclusion assuming that K, L_1, L_2 are complete.

Let $\alpha \in \mathcal{O}_1$ (the DVR of L_1) be such that its reduction modulo the maximal ideal generates the residue field extension over K. Then $K(\alpha) = L_1$, because $[K(\alpha) : K]$ is at least the residual degree $f(L_1|K)$, which in turn equals $[L_1 : K]$ since this last extension is unramified.

Then $L = L_2(\alpha)$. Let g be the minimal polynomial of α over L_2; its coefficients are in $\mathcal{O}_2$ (because α is integral over $\mathcal{O}$). Also, the reduction $\tilde{g}$ of g modulo $\mathbb{P}_2$ has not repeated roots (since the reduction of α is separable modulo $\mathcal{P}$). Therefore, by Proposition 2.6.3 $\tilde{g}$ must be irreducible. Hence the residual degree of L/L_2 is the same as the degree, proving that L/L_2 is unramified. That L/K is unramified now follows from the multiplicativity of the ramification indices. □

The proof simplifies when the residual degrees are 1 (which holds for function fields over an algebraically closed field); in this case the result is immediate once we have reduced to the complete case.

Exercise 2.6.7.

(i) Let Ω_p be an algebraic closure of $\mathbb{Q}_p$. Prove that Ω_p is not complete w.r. to the (unique by Corollary 2.6.9) extended valuation. Also, prove that the completion of Ω_p is algebraically closed. (See [10].)

(ii) **(Krasner's Lemma** [1], [18]**)** let L/K be a Galois extension of the complete field K. Let $y, z \in L$ and assume that $|z - y| < \min_{\sigma(y) \neq y} |y - \sigma(y)|$, where $\sigma(y)$ are the conjugates of y over K. Prove that $K(y) \subset K(z)$.

(Hint: otherwise there exists $\sigma \in Gal(L/K)$ fixing z and moving y; by Corollary 2.6.9, $|\sigma(y) - z| = |y - z|$, which is impossible.)[21]

---

[21] Note that this need not be true in the inseparable case. For instance let char$(k) = p > 0$ and let $K = k((x))$. Then $Y^p - x$ is irreducible over K and has a single root y, so the assumption is verified with $z = 0$; but $y \notin K$.

Deduce that if $f \in K[X]$ is monic and irreducible over K and $g \in K[X]$ is monic, very close to f and $\deg g = \deg f$, then g is irreducible.

(Hint: look at a root of g close to a root of f: continuity of the roots of a polynomial.)

(iii) Let K have characteristic 0 and let $\hat{K}$ be its completion w.r. to a DVR of K. Also, let E be a finite extension of $\hat{K}$. Prove that there exists a finite extension L/K such that E is the closure of L. [22]

(iv) Prove that $\mathbb{Q}_p$ has only finitely many extensions of given degree.

(Hint: use (ii).) Extend the result to other complete fields w.r. to a DVR.

As a complement to this assertion, provide details in the following construction of infinitely many separable extensions of degree p of $k((x))$, with k an algebraic closure of $\mathbb{F}_p$: For $c \in k$, let $L_c := k((x))(y_c)$ where $y_c^p - y_c = c/x$. Prove that the $L_c/k((x))$ are separable, totally ramified, of degree p, and Galois with group $y_c \mapsto y_c + l, l \in \mathbb{F}_p$. Suppose $L_c = L_1$, say. Using e.g. the Galois group, prove that then $y_c = a(x) + by_1$ where $a \in k((x))$, $b \in \mathbb{F}_p^*$, and conclude that necessarily $c \in \mathbb{F}_p$.

(v) Let K be a p-adic field (=finite extension of $\mathbb{Q}_p$). Prove that any automorphism of K is continuous.

(Hint: find an algebraic characterization of neighborhoods of 1. For instance, note that if $q = p^f$ is the order of the residue field, an element $x \in K$ is congruent to 1 modulo $\mathcal{P}$ if and only if it may be written as $y^{m(q-1)}$ for $y = y_m \in K$ and arbitrarily large exponents $m \in \mathbb{N}$; use Hensel's lemma.)

Extend the result to other complete fields w.r. to a DVR.[23]

(vi) Prove that there does not exist a field F properly contained in $\mathbb{Q}_p$ and such that $[\mathbb{Q}_p : F]$ is finite.

(Hint: Let K be a Galois closure of $\mathbb{Q}_p/F$. Then K is complete and since F is the fixed field it is closed by (v) above.)

(vii) Let K be a field complete w.r. to a DVR $(\mathcal{O}, \mathcal{P})$ and let F be a finite separable extension of $\mathcal{O}/\mathcal{P}$. Prove that (in a given algebraic closure of K) there exists a unique separable and unramified extension L/K such that the extension of residue fields is $F/(\mathcal{O}/\mathcal{P})$.

---

[22] Again, it may be proved that this need not be true in positive characteristic.

[23] These results may be seen as analogues to the well-known fact that every automorphism of $\mathbb{R}$ is continuous. In turn, this follows just by observing that any automorphism preserves the ordering, since it sends squares in squares, and these last are the positive elements.

(Hint: lift with Hensel's lemma a generator $\tilde{x}$ for $F/(\mathcal{O}/\mathcal{P})$, by lifting its minimal polynomial over $\mathcal{O}/\mathcal{P}$.)

Prove that in the case of number fields this extension is Galois.

(Hint: the residue field extension is Galois.)

(viii) Let $f \in \mathbb{Z}[x]$ be a given polynomial and let p be a given prime number. Find an effective algorithm to find the degrees of its irreducible factors over $\mathbb{Z}_p$.

(Hint: First, detect multiple factors of f by Euclid Algorithm, to reduce to the case when f has no such factors. Then, note that we have an equation $a(x)f(x) + b(x)f'(x) = D \neq 0$ over $\mathbb{Z}[x]$. Factor f modulo p^n where n is large enough w.r. to D, and then mimic Hensel's lemma to prove that a factorization may be lifted to $\mathbb{Z}$.)

(ix) Notation as in (viii), prove that if f is fixed, there exist infinitely many prime numbers p such that f has a linear factor over $\mathbb{Q}_p$.

(Hint: look at prime divisors of the numbers $f(n), n \in \mathbb{N}$ and apply the previous lifting.)

Moreover prove that for infinitely many primes f splits into linear factors over $\mathbb{Q}_p$.

(Hint: Apply the previous fact to a minimal polynomial of a a generator of a splitting field for f over $\mathbb{Q}$.)

(x) Let $L/\mathbb{Q}$ be a finitely generated field. Prove that, for all primes p, L may be embedded in a finite extension of $\mathbb{Q}_p$.

(Hint: $\mathbb{Q}_p$ has infinite transcendence degree.)

Prove also that for infinitely many primes p, L may be embedded in $\mathbb{Q}_p$.

(Hint: Use an argument as in (ix).)

(xi) Let L/K be a finite extension, assuming K complete w.r. to a DVR $\mathcal{O}$, and let $\mathcal{D}$ be the valuation ring of L. Assume also that the residue field extension is separable. Prove that $\mathcal{D} = \mathcal{O}[u]$ for a suitable $u \in \mathcal{D}$.

(Hint: Let π be a local parameter of $\mathcal{D}$ and let $t \in L$ such that its reduction generates the residue field extension. Then $\mathcal{D}$ is generated over $\mathcal{O}$ by the products $t^a \pi^b$. Let $f \in \mathcal{O}[X]$ be a monic polynomial such that its reduction mod $\mathcal{P}_\mathcal{D}$ is a minimal equation for t. Prove that either $f(t)$ or $f(t+\pi)$ is a local parameter for $\mathcal{D}$ and conclude by choosing either $\pi = f(t)$ or by changing t to $t + \pi$ and π to $f(t + \pi)$.)

(xii) Let K be complete w.r. to a compact DVR. Let $s(x) \in K[[x]]$ be a formal power series which converges in some disk around 0 in $\mathcal{O}$. Prove that if s has infinitely many zeros in that disk, it is identically zero.

(Hint: Expand around a limit point, by rearranging the series.)

(xiii) Let K be a finite extension of $\mathbb{Q}_p$, with DVR $(\mathcal{O}, \mathcal{P})$. Prove that the familiar series for $\exp(x)$ and $\log(1 + x)$ have a positive radius of convergence. Prove also that in the appropriate disk, $\exp(\log(1 + x)) = 1 + x$.

Let now $a \in \mathcal{O}^*$. Prove that there exists a positive integer h such that $a^{p^h(p-1)x}$ is defined by a convergent power series for $x \in \mathbb{Z}_p$. (Hint: Observe that $a^{p^h(p-1)} \equiv 1$ modulo high powers of $\mathcal{P}$ if h is large, so one may take its log and then use $a^{xp^h(p-1)} = \exp(x \log(a^{p^h(p-1)}))$.)

(xiv) (**Theorem of Skolem-Mahler-Lech**) Let $a_1, \ldots, a_r \in \overline{\mathbb{Q}}, b_1, \ldots, b_r \in \overline{\mathbb{Q}}[T]$. Prove that the set of zeros of the function $\sum_{i=1}^r b_i(n)a_i^n$, $n \in \mathbb{N}$, is a finite union of arithmetic progressions plus a finite set. (Hint: Embed into a p-adic field K a number field containing the stated (finitely many) quantities, in such a way that the a_i are in $\mathcal{O}^*$. Then partition $\mathbb{N}$ into a finite number of progressions modulo $p^h(p - 1)$ where h is large enough so to apply (xiii) with each a_i in place of a. On each progression our function is represented by a power series, and by (xii) cannot have infinitely many zeros in $\mathbb{Z}_p$ unless it vanishes identically there.)

The functions $f(n)$ of the stated shape are relevant for several reasons; they represent the 'general' solutions of linear recurrent sequences $\sum_{i=1}^l a_i f(n + i) = 0$ (a_i algebraic numbers not all 0) and also the coefficient of rational power series $R(x) = \sum_{n=0}^\infty f(n)x^n$.

(xv) (**Weierstrass preparation**) Let $\mathcal{O}$ be a complete DVR and let $s(x) = s_0 + s_1 x + \ldots \in \mathcal{O}[[x]]$. Suppose that $v(s_r) = 0$ but $v(s_i) > 0$ for $0 \le i \le r - 1$. Prove that there exists a monic polynomial $p \in \mathcal{O}[x]$ of degree r and an invertible series $t(x) \in (\mathcal{O}[[x]])$ such that $s(x) = p(x)t(x)$. Prove also that they are unique. (Hint: We may write $s(x) = cx^r + \pi q(x) + O(x^{r+1})$ where $c \in \mathcal{O}^*$, $v(\pi) = 1$ and q is a polynomial of degree $< r$. This yields $x^r = s(x)u(x) + \pi q_1(x) + \pi x^r u_1(x)$ where $u, u_1 \in \mathcal{O}[[x]]$ and q_1 is as q. Use this equation to substitute for x^r in the last term of this same equation; then iterate this procedure and use finally the completeness of $\mathcal{O}$.)

Use this result to obtain another proof of (xii).

(xvi) (**Newton polygon**) For K a field with DVR, let $f(x) = a_0 + a_1 x + \ldots + a_d x^d \in K[x]$ with a root $\rho \in K$. Consider the convex closure of the set $(0, +\infty)$, $(0, -v(a_0)), \ldots, (d, -v(a_d))$. Prove that $v(\rho)$ lies among the slopes of the sides of this polygon.[24]

(xvii) Let K be complete w.r. to a DVR $\mathcal{O}$ and let L/K be a finite totally ramified separable extension. prove that $L = K(\alpha)$ where α is the

---

[24] More precise conclusions can be proved: see [1, 10].

root of an Eisenstein polynomial (see Exercise 2.5.2(vi)) of degree $[L : K]$.

(Hint: Let α be a local parameter and use Corollary 2.6.9.)

(xviii) (**Artin-Hasse series**) Let p be a prime number and consider the formal power series over $\mathbb{Q}_p$ say: $E(x) := \exp(x + \frac{x^p}{p} + \frac{x^{p^2}}{p^2} + \ldots)$. Prove that $E(x) \in \mathbb{Z}_p[[x]]$.

(Hint: Observe the identity $E(x)^p = \exp(px)E(x^p)$ and argue by induction on the coefficients $a_0, a_1, \ldots$, setting $E(x) = \sum_{i=0}^{\infty} a_i x^i$.)

Prove that $y = E(x)$ satisfies the differential equation $xy' = zy$, where $z = x + x^p + x^{p^2} + \ldots$. Prove that z does not represent an algebraic function of x (hence the same is true for $E(x)$). (Note that z it is too 'lacunary': as we shall see in the appendices to Ch. III, the 'gaps' in an algebraic power series in 0 charact. are bounded; alternatively, it satisfies $z(x^p) = z(x) + x$, which may be shown to prevent algebraicity; prove this by looking at minimal polynomials.) Prove that its reduction modulo p does: it satisfies $z^p - z + x = 0$. Then the reduction of $E(x)$ may be expressed a series $e(z) \in k[[z]]$, satisfying $e'(z) = -e(z)/(z^{p-1} - 1)$.

Let now l be a prime number $\neq p$. Prove that $\prod_{\theta^l=1} E(\theta x) = 1$. Deduce that $E(x)$ does not represent an algebraic function of x, even in characteristic p. (Look at the values of the rational function x at the zeros ζ and poles π of $E(x)$, in the function field $k(x, E(x))$; if for no $\zeta, \eta, x(\zeta)/x(\pi)$ is an l-th root of 1 we have a contradiction.) It is also possible to show that $E(x) = \prod_{m=1}^{\infty} (1 - x^m)^{\frac{\mu(pm)}{m}}$, where μ is the Möbius function (see [5] and also [10]). In particular, this identity provides alternative proofs of some of the above assertions.

(xix) Let K be a proper subfield of $\mathbb{Q}_p$. Prove that $\mathbb{Q}_p/K$ cannot be purely trascendental.

(Hint: letting x be any element in a transcendence basis for $\mathbb{Q}_p/K$, let $a \in \mathbb{Z}$ so that ax is a square in $\mathbb{Q}_p$.)

2.7. Notes to Chapter 2

Section 2.1 The theory of VRs was initiated by Krull, and the viewpoint of algebraic curves and algebraic functions as depending on DVRs of function-field is mainly due to Chevalley. In this presentation we have followed heavily his book [7] at many points, and also other sources, like Lang's [16] and Serre's [24].

Section 2.2. We have followed essentially [7] in the proof of Theorem 2.2.1 (and also in some applications), with a little additional generality. The result in Exercise 2.2.2 is a special case of a result by Kronecker, which may also be proved in a more explicit way: see Theorem 11, p. 59 of Schinzel's book [31].

Section 2.3. One can consider absolute values on a field other than those coming from some DVRs (as for instance in the case of number fields). Especially for ultrametric absolute values, several results continue to hold; see [1, 10, 21].

Section 2.4. The approximation theorem 2.4.2 is due to Artin and Whaples; it may be seen as an approximation theorem for several points in $\mathbb{P}_1$. See [20] for an extension to arbitrary curves (it holds only after suitable extensions of the ground field).

Section 2.5. We have followed [7] for Theorem 2.5.1, introducing some modifications and additional generality. We have omitted the important notion of *tame ramification* which occurs when the characteristic does not divide the ramification index. Apart from Exercise 2.5.2(iv), we have also omitted the Galois theory of extensions of DVRs. When L/K is finite Galois, with group G, one can define the *decomposition group* $G_{\mathcal{D}}$ of a DVR $\mathcal{D}$ in L above $\mathcal{O}$ as the subgroup of G stabilizing $\mathcal{D}$ as a set. One can also define a filtered sequence of *ramification (sub)groups* (of $G_{\mathcal{D}}$), the first of which is the *inertia group* $G_{\mathcal{D}}^{(0)} = \{\sigma \in G_{\mathcal{D}} : \sigma(x) \equiv x$ (mod $\mathcal{D}$), $\forall x \in \mathcal{D}\}$. One can prove that $G_{\mathcal{D}}/G_{\mathcal{D}}^{(0)}$ is isomorphic to the Galois group of the residue extension. See *e.g.* [24] for a detailed and vast account.

Section 2.6. There are accounts of the theory of completions in [1, 6, 18, 24], and also in [7] for the function-field case.

There exist several versions of Hensel's lemma, more or less general. In several variables, we may for instance state roughly the following generalization: *a nonsingular point on a the reduction modulo $\mathcal{P}$ of algebraic variety may be lifted to a point of the variety, over the complete ring $\mathcal{O}$.*

For equations in several variables, say $f(x_1, \ldots, x_n) = 0$, $f \in \mathbb{Z}[\mathbf{x}]$, to be solved in a p-adic field, there exist "effective" algorithms for the existence of solutions (based implicitly on Krasner's lemma). See P. J. COHEN, *Decision procedures for real and p-adic fields*, Comm. Pure Appl. Math., 1969.

Appendix A
Hilbert's Nullstellensatz

A.1. Generalities

The *Nullstellensatz*, or 'Theorem of the zeros'[1], found and proved by D. Hilbert in 1893, concerns the *common zeros* for all the polynomials in a subset I of $A := k[\mathbf{X}] := k[X_1, \ldots, X_n]$ (X_i indeterminates, *i.e.* algebraically independent over the field k).

In a way, it is a **criterion for the solvability of a system of algebraic equations**: $f(\xi_1, \ldots, \xi_n) = 0$ for all $f \in I$, where we require that $\xi := (\xi_1, \ldots, \xi_n) \in k^n$. It is usually assumed that k is algebraically closed (to ensure at least the solvability of equations of positive degree in a single variable).

Plainly, an identity of the type $\sum_{f \in I} a_f(X_1, \ldots, X_n) f(X_1, \ldots, X_n) = 1$, for elements $a_f \in A$, all but finitely many ones being zero, prevents the existence of common solutions.

It is on the other hand clear that for our purposes we may replace the set I with the ideal generated by it in A; so we may assume for the very beginning that I is an ideal. Then, the above identity just means that $I = A$.

The converse assertion (namely, that $I = A$ is the only obstruction to solutions) is just the content of the Theorem of the zeros, more precisely of what is usually called its 'weak form'.

Let us start by stating the theorem, and then, before proving it, pause for some remarks and for the deduction of some consequences.

In the sequel, let $\bar{k}$ be an algebraic closure of k.

Theorem A.1.1 (Nullstellensatz, weak form). *Let* $I \subset k[X_1, \ldots, X_n]$ *be a proper ideal. Then there exists a point* $P = (\xi_1, \ldots, \xi_n) \in \bar{k}^n$ *such that* $f(P) = 0$ *for all* $f \in I$.

---

[1] Literally: Theorem of the locus of zeros.

Note as above that that it suffices that $f(P) = 0$ for a set of generators of I. By Hilbert's Basis Theorem (see [2] or [17]) there exists such a set which is finite.

We also note at once the following consequence, which may be also seen as an alternative formulation:

Corollary A.1.2. *If k is algebraically closed, an ideal M of $k[X_1, ..., X_n]$ is maximal if and only if there exist $\xi_1, \ldots, \xi_n \in k$ such that M is generated by the $X_i - \xi_i$, $i = 1, \ldots, n$.*

Indeed, it is clear that every such ideal as in the corollary is maximal. To deduce the other half of this corollary, we may apply the Nullstellensatz A.1.1 with $I = M$. If $P = (\xi_1, \ldots, \xi_n)$ is as in its conclusion, we have that M is contained in the ideal generated by the $X_i - \xi_i$ (because this last is the ideal of the polynomials vanishing at P). Since M is maximal, the assertion follows.

Conversely, it is clear that Corollary A.1.2 implies directly the Nullstellensatz A.1.1, at least for algebraically closed k: it suffices to take a maximal ideal M containing I.

We shall state in a moment, after a few preliminaries, the so called 'strong form'; we shall see that the 'weak form' is a special case of it, but reciprocally we shall rapidly deduce the strong form from the weak one. Subsequently we shall prove the latter, illustrating a few possible methods.

A.1.1. Preliminaries on algebraic sets

For some readers' convenience, we briefly repeat here a few definitions already given in Chapter 1.

If I is an ideal of the polynomial ring A, we may consider the associated *affine algebraic set* V_I, that is the set $V_I := \{P \in k^n : f(P) = 0, \forall f \in I\}$ (also called 'zero-set of I'; see Section 1.3 of Chapter 1).

Conversely, it is possible to associate to a set $W \subset k^n$ an ideal $J = J_W := \{f \in A : f(P) = 0, \forall P \in W\}$. (As a matter of notation in the sequel, if we want to make explicit reference to the field k, we can use the notation $W(k)$ for the points of an algebraic set W with coordinates in k.)

If $W \subset k^n$ is an (affine) algebraic set, *i.e.* if W is defined as the zero set of some set of polynomials it is easy to see that $V_{J_W} = W$.

The reciprocal behaviour of this association is more delicate. We note at once that, if we start from an ideal I and we form V_I, the ideal J_{V_I} contains at least the *radical* $\sqrt{I}$ of I, namely the set of polynomials having some positive power in I: $\sqrt{I} := \{ f \in A :$ for some integer $n = n_f > 0,$ $f^n \in I \}$. The strong form of the Nullstellensatz states that there is equality if k is algebraically closed:

Theorem A.1.3 (Nullstellensatz, strong form). *Let k be algebraically closed, let $I \subset k[X_1, \ldots, X_n]$ be an ideal and let $W := V_I$ be the associated algebraic set. Then $J_W = \sqrt{I}$.*

This result is fundamental and unavoidable in Algebraic Geometry. As a simple example, let us see how it makes precise the equivalence (briefly stated in Chapter 1) between systems of generators $(x_1, \ldots, x_n)$ for a function field $K = k(x_1, \ldots, x_n)$ and *geometric affine models of K*. Let us recall how to obtain such a model V. We consider first the ideal I of relations among the x_i: $I = \{ f \in k[\mathbf{X}] : f(x_1, \ldots, x_n) = 0 \}$ and we put $V := V_I$ (as above).

Now we can consider the ideal J_V of V, defined a few lines above. In view of the 'strong form' we have $J_V = \sqrt{I}$ and since I is prime (because $k[x_1, \ldots, x_n] \subset K$ is a domain) we have $J_V = I$. The field K is then recovered as 'function field of the model V' (which is by definition the fraction field of $k[\mathbf{X}]/J_V$).

Exercise A.1.1.

(i) Prove directly the 'strong form' when $n = 1$ and also when the ideal I is generated by linear polynomials.

(ii) Prove directly (on eliminating a variable by means of a 'Bezout equation'= euclidean algorithm) the 'weak form' for $n = 2$, when I is generated by two polynomials f, g.

(Hint: it can be convenient to perform a linear substitution so to assume f, g monic in one of the variables.)

(iii) Let $g_1, \ldots, g_s \in \mathbb{Z}[\mathbf{X}]$. Suppose that for infinitely many prime numbers p the simultaneous congruences $g_i(a_1, \ldots, a_n) \equiv 0 \pmod{p}$, $i = 1, \ldots, s$, have a solution in integers $a_1, \ldots, a_n$ (dependent on p). Prove that the g_i have a common zero in $\overline{\mathbb{Q}}^n$.

Prove also the converse statement. (Hint for the converse statement: Let $\mathbb{Q}(\alpha)$ be a number field of definition for a common zero; find primes p such that a DVR above p in K has residual degree 1; in order to do this look at prime divisors for the values $f(n)$ at integers n of a minimal polynomial $f \in \mathbb{Z}[x]$ for α; compare with Exercise 2.6.7(viii)).

It is a rather deeper fact that in such a converse statement 'infinitely many' can be substituted by 'all but a finite number of primes', assuming however the stronger hypothesis that the algebraic set determined by the g_i is nonempty and 'geometrically irreducible'; this holds if and only if the g_i generate over $\bar{\mathbb{Q}}$ an ideal with prime radical.

(iv) Let $g_1, \ldots, g_s \in \mathbb{Z}[\mathbf{X}]$ be such that for every prime number p the reduced polynomials $\bar{g}_1, \ldots, \bar{g}_s \in \mathbb{F}_p[\mathbf{X}]$ have no common zero over an algebraic closure of $\mathbb{F}_p$. Prove that the g_i generate the unit ideal of $\mathbb{Z}[\mathbf{X}]$.

(Hint: Apply the Nullstellensatz to prove that the g_i generate the unit ideal over $\mathbb{Q}$ and over all the $\mathbb{F}_p$.)

(v) Let k be a field, L an extension of k, and let $f_1, \ldots, f_r \in k[\mathbf{X}]$ be polynomials generating the unit ideal in $L[\mathbf{X}]$. Prove that the f_i generate the unit ideal in $k[\mathbf{X}]$. (See the proof (A) below.)

(vi) If k, L are as in (v), prove directly that $f_1, \ldots, f_r \in k[\mathbf{X}]$ generate a maximal ideal in $k[\mathbf{X}]$ if and only if they generate a maximal ideal in $L[\mathbf{X}]$.

(vii) (**Ostrowski Theorem**) Let $g \in \mathbb{Q}[\mathbf{X}]$ be an absolutely irreducible polynomial (*i.e.* irreducible over the algebraic numbers). Prove that for all but finitely many primes p the reduction of g modulo p is defined and absolutely irreducible.

(Hint: The fact that g is irreducible results in a certain system of affine equations involving the coefficients of g to have no solutions. Hence the polynomials corresponding to such equations generate the unit ideal. Now, reduce modulo p an identity expressing this fact.) This result remains true over number fields, on taking the reduction modulo prime ideals of the ring of integers; the proof is the same.

(viii) (**Projective Nullstellensatz**) Let $f_1, \ldots, f_r \in k[\mathbf{X}]$ be homogeneous polynomials over the algebraically closed field k, having the origin as their only common zero. Prove that there exists a positive integer m such that the ideal generated by the f_i contains X_j^m, for every $j = 1, \ldots, n$.

A.2. Proof of the double implication 'weak form' $\Leftrightarrow$ 'strong form'

(A) To start with, let us show that the weak form is a special case of the strong form. So, let us suppose that the algebraic set W associated to I (over $\bar{k}$) is empty, *i.e.* that the polynomials in I have no common zero in k^n. Then (using the notions with $\bar{k}$ in place of k) we shall have $J_W = J_\emptyset = \bar{k}[\mathbf{X}]$. Therefore, by the 'strong form' (applied to $\bar{k}$) we should have $\sqrt{\bar{k}I} = J_W = \bar{k}[\mathbf{X}]$, so $1 \in \bar{k}I$, namely 1 lies in the ideal generated by I in $\bar{k}[\mathbf{X}]$; in other words, an equation $\sum_{f \in I} a_f(\mathbf{X}) f(\mathbf{X}) = 1$, $a_f \in \bar{k}[\mathbf{X}]$, holds. Now, let $1 = \omega_1, \omega_2, \ldots, \omega_m$ be a basis for the vector space over k generated by the nonzero coefficients of the a_f (finite dimension). We shall then be able to write each a_f as a linear combination $\sum_{i=1}^{m} \omega_i b_{i,f}(\mathbf{X})$ with $b_{i,f} \in k[\mathbf{X}]$. Then, since $I \subset k[\mathbf{X}]$, we have $\sum_{f \in I} b_{1,f}(\mathbf{X}) f(\mathbf{X}) = 1$, as required (see also Exercise A.1.1(v)).

(B) The proof of the converse implication is more subtle, but 'Rabinovitch trick' makes it simple. Assumptions and notations being as in the statement of the strong form, let $g \in J_W$, $g \neq 0$. We have to show that some power of g lies in I.

Let us consider a supplementary variable Y and the ideal $\tilde{I}$ in $k[\mathbf{X}, Y]$ generated by I and by $Yg(\mathbf{X}) - 1$. Plainly, the algebraic set W' in k^{n+1} associated to $\tilde{I}$ is empty: if $P = (\xi_1, \ldots, \xi_{n+1}) \in W'$, then on the one hand we have $f(P) = 0$ for all $f \in I$, hence $g(P) = 0$ (because $g \in J_W$). On the other hand, we have $\xi_{n+1}g(P) - 1 = 0$, a contradiction. Hence, on applying the 'weak form' to W' we obtain an identity $\sum_{f \in I} a_f(\mathbf{X}, Y) f(\mathbf{X}) + b(\mathbf{X}, Y)(Yg(\mathbf{X}) - 1) = 1$. Putting $Y = 1/g(\mathbf{X})$ we find $\sum_{f \in I} a_f(\mathbf{X}, 1/g) f(\mathbf{X}) = 1$, and multiplying by a sufficiently large power of g to render polynomials all the 'coefficients' $a_f(\mathbf{X}, 1/g)g^N$, we see that $g^N \in I$, namely $g \in \sqrt{I}$, as required.

Remark A.2.1.

(i) Argument (A) also shows that in the weak form of the result it is sufficient to assume k algebraically closed.

(ii) The trick in question is 'inspired' by the general fact that, for an algebraic set $Z \subset k^n$ and $g \in k[X_1, \ldots, X_n]$, the *open* variety $Z \setminus \{g = 0\}$ may be embedded as a *closed* algebraic set in k^{n+1}, just by considering the auxiliary equation $Yg = 1$.

In what follows we shall present some proofs for the 'weak form'. Some classical arguments use elimination theory, substantially to reduce to the simple case of one single variable (see *e.g.* [19]). On the contrary, here we shall follow different paths (see also [2, 17] for further approaches).

In view of Remark A.2.1(i), we shall suppose in the sequel of this appendix that k is algebraically closed.

We shall start with the classical case, which occurs when $k = \mathbb{C}$; however, this case is not fully representative, because a special proof is possible for it, easier than for the general case; this is due to the fact that $\mathbb{C}$ has infinite transcendence degree over $\mathbb{Q}$. In fact, such property implies the following useful:

Proposition A.2.1. *Let F be a subfield of $\mathbb{C}$ having finite transcendence degree over $\mathbb{Q}$ and let K/F be a finitely generated extension. Then K/F may be embedded in $\mathbb{C}$.*

Proof. We need to show that there exists an isomorphism $\varphi : K \to \mathbb{C}$ which equals the identity on F. Let us write $K = F(y_1, \ldots, y_m)$ and let us argue by induction on m, the case $m = 0$ being given by the assumption. In view of the inductive hypothesis, on replacing F by $F(y_2, \ldots, y_m)$, we may thus assume $m = 1$. If y_1 is transcendental over F, let $\Pi \in \mathbb{C}$ be transcendental over F: such a Π exists because $F/\mathbb{Q}$ has finite transcendence degree, while $\mathbb{C}/\mathbb{Q}$ has infinite (and even non-denumerable) transcendence degree (a well-known fact that the interested reader will easily prove). Now, define $\varphi(y_1) := \Pi$; it is then immediate to verify that an isomorphim as required may be obtained on setting $\varphi(r(y_1)) := r(\Pi)$ for $r \in F(y_1)$.

If on the other hand y_1 is algebraic over F, let $P \in F[X]$ be its minimal polynomial, and let $\rho \in \mathbb{C}$ be a root of P. It is again easy to show that the definition $\varphi(y_1) = \rho$ extends to an isomorphism of $F(y_1)$ onto $F(\rho) \subset \mathbb{C}$. $\qquad\square$

Remark A.2.2. In Algebraic Geometry the proposition is often applied to reduce the case of a general field of characteristic zero to the case of $\mathbb{C}$. This may be done when the relevant statements involve only finitely many polynomials and coefficients. Such reduction is often referred to as the 'Lefschetz principle'. For instance, it allows one to use of tools from topology for treating certain statements of purely algebraic nature.

Proof of the 'weak form' A.1.1 for $k = \mathbb{C}$. Let M be a maximal ideal of $\mathbb{C}[\mathbf{X}]$ containing I. By Hilbert's basis theorem, M is finitely generated. Let then F be the field generated over $\mathbb{Q}$ by the coefficients of a finite set of generators for M. The ideal $M' := M \cap F[\mathbf{X}]$ of $F[\mathbf{X}]$ is certainly prime[2] and so $D := F[\mathbf{X}]/M'$ is a domain. By Proposition A.2.1, its fraction field may be embedded in $\mathbb{C}$ (over F); let then ξ_i be a complex number corresponding to the class of X_i in D, in this embedding of D in $\mathbb{C}$. Since every $f \in F[\mathbf{X}]$ in the above mentioned set of generators for M lies in M', we have $f(\xi_1, \ldots, \xi_n) = 0$. Hence $(\xi_1, \ldots, \xi_n)$ is a zero of M (and *a fortiori* of I) in $\mathbb{C}^n$. $\qquad\square$

As alluded above, this argument does not express the whole content of the Nullstellensatz, since it uses a rather peculiar property of the ground field $\mathbb{C}$. Let us see another proof, this time fully general, which follows immediately from the properties of VRs that we have studied.

---

[2] In fact, it is maximal, see Exercise A.1.1(vi) above.

First general proof of the 'weak form A.1.1, *for algebraically closed k.*
Let I be as in the statement, so that there exists a maximal ideal M of
$k[\mathbf{X}]$ containing I. Then $K := k[\mathbf{X}]/M$ is an algebra over k [3] which is
finitely generated and moreover a field. By Application F of Theorem
2.2.1, we have that K/k is a finite extension, and therefore $K = k$. If
$\xi_i \in k$ is the image of X_i modulo M, we see that $f(\xi_1, \ldots, \xi_n) = 0$ for
all $f \in I$ (actually for all $f \in M$), and the conclusion follows.   □

We go on with another proof of the weak form. The main idea for
this proof is substantially that (by the simple 'Theorem of the primitive element') an *(affine) algebraic variety is birationally equivalent to a
hypersurface*; on the other hand for hypersurfaces the Nullstellensatz is
very easy. For simplicity we shall suppose char$(k) = 0$, but the argument
may be modified for the general case.

Second general proof of the 'weak form' A.1.1. As mentioned, we assume that k is algebraically closed of characteristic zero. Moreover, as
above we can also assume that I is a prime (and in fact maximal) ideal.
    Let $D := k[\mathbf{X}]/I$ and let K be the fraction field of D. If $z_1, \ldots, z_r$
is a transcendence basis for K/k and if $L := k(z_1, \ldots, z_r)$, we find that
K/L is an algebraic and finitely generated extension, and thus a finite
extension. By the Theorem of the primitive element (see, *e.g.*, [17] - it is
here that we are using the zero-characteristic) we may write $K = L(u)$
for a suitable $u \in K$, algebraic over L. Let $P(z_1, \ldots, z_r, u) = 0$ be a
minimal equation for u over $k[z_1, \ldots, z_r]$.[4] It will have positive degree
in u. Let now x_i be the image of X_i in D; then (since $L(u) = L[u]$) we
may write

$$x_i = \frac{Q_i(z_1, \ldots, z_r, u)}{\Delta(z_1, \ldots, z_r)}$$

for suitable polynomials (over k) Q_i and $\Delta \neq 0$.
    Let $f_1, \ldots, f_h$ be (a finite system of) generators for I.
We have $f_j(x_1, \ldots, x_n) = 0$ and thus we find that the rational func-
tions $f_j(Q_1(\mathbf{z}, U)/\Delta(\mathbf{z}), \ldots, Q_n(\mathbf{z}, U)/\Delta(\mathbf{z}))$ (in the independent variables $\mathbf{z} = (z_1, \ldots, z_r)$ and U) vanish for $U = u$; therefore their respect-
ive numerators are divisible by $P(\mathbf{z}, U)$, whereas all their denominators
divide a suitable power of $\Delta(\mathbf{z})$.

---

[3] Note that k embeds isomorphically in the quotient ring, because M is a proper ideal.

[4] By this we mean an equation of minimal degree and with coprime coefficients in $k[z_1, \ldots, z_r]$.

Let now $(\eta_1, \ldots \eta_r, \mu) \in k^{r+1}$ be such that $P(\eta_1, \ldots, \eta_r, \mu) = 0$ and $\Delta(\eta_1, \ldots, \eta_r) \neq 0$. Such a point exists: it suffices to take $(\eta_1, \ldots, \eta_r) \in k^r$ which is not a zero of the product of Δ by the coefficients of P (viewing P as a polynomial in U over $k(\mathbf{z})$), and successively to find a root $U = \mu \in k$ of $P(\eta_1, \ldots, \eta_r, U)$.

Let us finally put $\xi_i = Q_i(\eta_1, \ldots, \eta_r, \mu)/\Delta(\eta_1, \ldots, \eta_r)$. Since $P(\mathbf{z}, U)$ divides all the numerators of the $f_j(Q_1(\mathbf{z}, U)/\Delta(\mathbf{z}), \ldots, Q_n(\mathbf{z}, U)/\Delta(\mathbf{z}))$, we see immediately that $(\xi_1, \ldots, \xi_n)$ is a zero of f_j (for every j) and thus is a zero of I in k^n. $\square$

We conclude this appendix with two important applications.

The first one proves what we asserted in Chapter 1.

Proposition A.2.2. *An (affine) variety of positive dimension over an algebraically closed field k has infinitely many points*

Recall that by *variety* we mean an irreducible algebraic set; equivalently, an affine algebraic set $V \subset k^n$ is a variety if the associated ideal J_V is prime. In such case we call *dimension* of V the number $\mathrm{trdeg}(K/k)$, where K is the fraction field of the domain $k[V] := k[\mathbf{X}]/J_V$.

We observe that a proof of the proposition above follows implicitly from the argument just given for the 'weak form'. However the assertion also follows directly from the 'strong form'.

Proof. Suppose that $\dim V > 0$ but that V is a finite set. Let $f \in k[V]$ be a transcendental element over k and let F be a representative for f in $k[\mathbf{X}]$. Set also $G(\mathbf{X}) := \prod_{P \in V}(F(\mathbf{X}) - f(P)) \in k[\mathbf{X}]$. For $P \in V$ we have $F(P) = f(P)$ and therefore $G(P) = 0$. Hence G vanishes on V and thus $G \in \sqrt{J_V} = J_V$ (the equality being true because J_V is prime). Then the reduction of G modulo J_V vanishes; but such reduction is $\prod_{P \in V}(f - f(P))$; hence f is algebraic over k, against the assumptions. $\square$

As to the second application, recall from Section 3 of Chapter 1 that a rational function f on an irreducible algebraic set V is called *regular* at a point $P \in V$ if f may be written as a quotient $A(\mathbf{x})/B(\mathbf{x})$ where $A, B \in k[V]$ and $B(P) \neq 0$. This notion is local and it is clear that the elements of $k[V]$ are globally regular, that is regular at every point. The Nullstellensatz allows us to recover $k[V]$ just as the subset of globally regular elements in $k(V)$:

Proposition A.2.3. *Let V be an irreducible affine variety over the algebraically closed field k and let $f \in k(V)$ be a rational function on V, regular at every point of $V(k)$. Then $f \in k[V]$.*

Proof. The assumption says that for every point $P \in V(k)$ we may write $f = A_P/B_P$ with $A_P, B_P \in k[V]$ and $B_P(P) \neq 0$. Let I be the ideal of $k[\mathbf{X}]$ generated by the ideal J_V of V and by all the B_P, for $P \in V(k)$. (We tacitly mean to choose a representative in $k[\mathbf{X}]$ for B_P: it makes no difference which representative.) If $Q \in k^n$ is a zero of I then $Q \in V(k)$ (because $I \supset J_V$) and moreover $B_P(Q) = 0$ for all P. This gives a contradiction because $B_Q(Q) \neq 0$ by assumption.

We conclude that such a Q does not exist, and so I has no zeros in k^n and by the 'weak form' it coincides with $k[\mathbf{X}]$. Therefore there is an identity $1 = S(\mathbf{X}) + \sum_{P \in V} C_P(\mathbf{X}) B_P(\mathbf{X})$, where $S \in J_V$ and $C_P \in k[\mathbf{X}]$ for all $P \in V$. Reducing modulo J_V and multiplying by f gives $f = \sum_{P \in V} C_P f B_P = \sum_{P \in V} C_P A_P \in k[V]$ as wanted. □

This argument reminds of, and may be viewed as, an algebraic analogue of the 'partitions of unity', often used in Differential Geometry to patch local informations, which actually is the same goal of the proof just given.

Appendix B
Puiseux series

B.1. Field of definition, convergence and Eisenstein Theorem

Let $f \in k[X, Y]$ be a polynomial of positive degree δ in Y, where k is a field of characteristic zero (with algebraic closure $\bar{k}$). Recall from Exercise 2.6.2 that the equation $f(x, Y) = 0$ has a formal solution $Y = y = y(x^{1/d}) \in \bar{k}((x^{1/d}))$ for some positive integer d.

In this appendix we want to study some features of this kind of series. But first let us briefly review the alluded argument, at the light of the whole Section 2.6 above: Let $K = k(x, y)$, where $f(x, y) = 0$. We can embed $k(x)$ into $k((x))$ and correspondingly we can embed K into a finite extension of $k((x))$. This is of the type $k((t))$, where t is a local parameter, and must be totally ramified, because of Theorem 2.6.8 and because the residual degree is 1. If $d = [k((t)) : k((x))]$ we have $x = t^d s(t)$ where $s \in k[[t]]$ does not vanish at 0. Extracting a d-th root $\rho(t)$ of $s(t)$ we see that we may replace t with $t\rho(t)$, and in fact take t so that $t^d = x$, which completes the argument, since then y lies in $k((t))$.

Actually, this approach also shows that f has a full set of series-solutions of that shape. In this appendix we will briefly discuss them from a more arithmetical viewpoint.

Suppose that d is a 'minimal denominator' for $y(x^{1/d})$, in the sense that $y(x^{1/d}) \notin k((x^{1/e})$ for any proper divisor e of d; this is equivalent with the fact that $y(t) \notin k((t^s))$ for a divisor $s > 1$ of d. Then plainly two facts occur:

(i) For distinct d-th roots of unity θ, θ', the series $y(\theta x^{1/d})$ and $y(\theta' x^{1/d})$ are distinct.

(ii) For $\theta^d = 1$, the series $y(\theta x^{1/d})$ is a solution of $f(x, Y) = 0$. (This is proved just by using the map $x^{1/d} \mapsto \theta x^{1/d}$ in the equation $f(x, y) = 0$, i.e. the substitution $x = t^d, t \mapsto \theta t$ in the identity $f(t^d, y(t)) = 0$.)

It is worth noticing that this last observation also implies that there is an element of order d in the Galois group of the polynomial f over $k((x))$.

These two remarks show that the polynomial

$$g(x, Y) := \prod_{\theta^d=1} (Y - y(\theta x^{1/d}))$$

divides $f(x, Y)$ in the ring $\bar{k}((x^{1/d}))[Y]$. However, since the coefficients of g are invariant under the maps $x^{1/d} \mapsto \theta x^{1/d}$, we see that actually $g \in \bar{k}((x))[Y]$. Moreover g is irreducible over $\bar{k}((x))$: indeed, if $y(x^{1/d})$ is a root of a factor, that factor must have all the roots $y(\theta x^{1/d})$ by invariance. But such roots are distinct by the minimality of d, and the conclusion follows. Then we see that this irreducible factor determines an extension of $\bar{k}((x))$ of degree d, corresponding to the given root of f. Since the residue-field extension is trivial (over $\bar{k}$), this extension is totally ramified, with ramification index d.

Reciprocally, all the irreducible factors of f over $\bar{k}((x))$ arise in this way, because f has a full set of Puiseux series roots (by Exercise 2.6.2 again). So, (by Section 2.6) the Puiseux series in fact determine the splitting of the standard DVR of $k((x))/k$ in an algebraic extension.

We add that in concrete cases it is effectively possible to compute the ramification indices, for instance by using the 'Newton polygon' (see [10] or [36]). However, we shall not pause on this here. Instead, we shall focus on some natural properties of the coefficients of such series. Two questions (at least) naturally arise:

Question 1. Do the coefficients of y actually all lie in a finite extension of k?

Question 2. Suppose that k is complete with respect to an absolute value $|\cdot|$ which satisfies the usual properties (this may be for instance the usual absolute value on $\mathbb{C}$ or an ultrametric absolute value coming from a DVR of k). Is the series for y necessarily convergent in some disk around 0?

And, if the answer is in the affirmative do the values of the series in its circle of convergence yield solutions of the corresponding algebraic equations?

Below we shall provide simple affirmative answers. Question 2 also leads to the problem of bounding below the radius of convergence (for instance by some explicit quantity associated to the coefficients of f). This is a subtle problem, especially in the ultrametric case (see [10]). We shall see that over number fields, where we have infinitely many ultrametric absolute values coming from the DVR, the radii of a given series are 'almost always' (*i.e.*, with only finitely many exceptions) $\geq 1.$[1] This will follow from a stronger property, discovered by Eisenstein: *The denominators of the coefficients are altogether divisible only by finitely many prime ideals of the ring of integers of* k. This means that for only finitely many DVRs $(\mathcal{O}, \mathcal{P})$ of k some coefficient may have a pole at $\mathcal{P}$; note that if $\mathcal{P}$ is not in this finite set the valuation of any coefficient is ≤ 1, whence the radius of convergence is ≥ 1.

Before going to the proofs we perform some simple normalizations. First, on changing x to x^d we may assume that $y \in \bar{k}((x))$. Secondly, we may multiply y by some power of x to assume that y lies in $\bar{k}[[x]]$ and satisfies $y(0) = 0$. We then write $y(x) = \sum_{m=1}^{\infty} a_m x^m$. We also put $y_n = y_n(x) = \sum_{m=1}^{n} a_m x^m$.

We start by answering Question 1:

Theorem B.1.1. *There exists a finite extension of k containing all the coefficients a_m. More precisely, the degree $[k(a_1, a_2, \ldots) : k]$ does not exceed $\deg_Y f$.*

Proof. Let L be a Galois closure of $k(a_1, a_2, \ldots)/k$, with (possibly infinite) Galois group G. Note that G acts on $y(x)$ by acting on the coefficients. Plainly, for $\sigma \in G$ we have $f(x, y^\sigma(x)) = f(x, y(x)) = 0$, since G fixes the coefficients of f. Hence all the conjugate series $y^\sigma(x)$ are roots of $f(x, Y) = 0$ in $k((x))$, and there can be at most $\deg_Y f$ of them. Thus, if G_1 is the stabilizer of $y(x)$, we have $[G : G_1] \leq \deg_Y f$. Let then K be the fixed field of G_1. We have $y(x) \in K[[x]]$ and also $[K : k] = [G : G_1] \leq \deg_Y f$, as required. $\qquad\square$

Remark B.1.1. Note that this need not be true in positive characteristic: if $t_0, t_1, \ldots$ are infinitely many indeterminates over a field L of characteristic p, then the series $y(x) = \sum_{i=0}^{\infty} t_i x^i$ satisfies $y^p = \sum_{i=0}^{\infty} t_i^p x^{pi} \in L(t_0^p, t_1^p, \ldots)[[x]]$; however the coefficients $t_0, t_1, \ldots$ of y generate an extension of $L(t_0^p, t_1^p, \ldots)$ of infinite degree.

Before continuing we establish some further facts. Enlarging k (to a finite extension if needed) we assume that $y \in k[[x]]$. In the sequel we shall denote by v the order function on $k((x))$ and by '$O(x^m)$' a power series of order $\geq m$. We may suppose that f has no multiple factors, so in particular, the partial derivative $f_Y(x, y(x))$ does not vanish.[2] We let $h \geq$

---

[1] Actually equal to 1 up to such exceptions, by Exercise B.1.2(vi) below.

[2] Here we are using that $\mathrm{char}(k) = 0$.

0 be its order, so h is a natural number and we may write $f_Y(x, y(x)) = bx^h + O(x^{h+1})$ for some $b \in k^*$. Plainly, for $n \geq h$ we shall also have $f_Y(x, y_n(x)) = bx^h + O(x^{h+1})$.

Exercise B.1.1. Prove that h is bounded in terms only of deg f.

Theorem B.1.2. *For any absolute value $|.|$ on k, the series $y(x)$ has a positive radius of convergence. That is, there exist reals $B_1, B_2 > 0$ such that $|a_m| \leq B_1 B_2^m$ for all $m \geq 1$.*[3]

Proof. We first establish a recurrence relation for the a_m. We fix an integer $n \geq h$. On the one hand we have $y_{n+1}(x) = y_n(x) + a_{n+1}x^{n+1}$ so, by Taylor expansion,

$$f(x, y_{n+1}(x)) = f(x, y_n(x)) + a_{n+1}x^{n+1} f_Y(x, y_n(x)) + O(x^{2n+2}),$$

which leads to

$$f(x, y_{n+1}(x)) = f(x, y_n(x)) + a_{n+1}bx^{n+1+h} + O(x^{n+h+2}).$$

On the other hand, $y_{n+1}(x) = y(x) + O(x^{n+2})$, whence, expanding again,

$$\begin{aligned} f(x, y_{n+1}(x)) &= f(x, y(x)) + f_Y(x, y(x))O(x^{n+2}) + O(x^{2n+4}) \\ &= O(x^{n+h+2}). \end{aligned}$$

Comparing the last two displayed equations we get

$$a_{n+1}x^{n+h+1} = -\frac{1}{b}f(x, y_n(x)) + O(x^{n+h+2}).$$

Hence we have that, for all $n \geq h$,

$$a_{n+1} = -\frac{1}{b} \cdot \text{coefficient of } x^{n+h+1} \text{ in } f(x, y_n(x)). \qquad \text{(B.1)}$$

In particular, by an easy inspection we derive the estimate:

$$|a_{n+1}| \leq c_1 \max_{0 \leq s \leq \delta} \left(1 + \sum_{j_1,\dots,j_s} |a_{j_1}| \cdots |a_{j_s}| \right), \qquad \text{(B.2)}$$

where $c_1 = c_1(f)$ and the summation is taken over all s-tuples $(j_1, \dots, j_s)$ of positive integers with $n + h + 1 - \delta \leq j_1 + \dots + j_s \leq n + h + 1$ and $1 \leq j_1, \dots, j_s \leq n$.

---

[3] The equivalence of these conclusions is standard for the usual (archimedean) absolute value on $\mathbb{C}$, and may be proved generally in the same way.

We assume from now on that $n \geq 2(\delta + h)$ and we let

$$A := \max(1, |a_1|, \ldots, |a_{2(\delta+h)}|).$$

We also let $c_1, c_2, \ldots$ be numbers > 1 depending only on f.

We are going to show by induction on $m \geq 2(\delta + h)$ that $|a_m| \leq \alpha B^m/m^2$ for suitably choosen positive 'small' α and 'large' $B > 4$. This bound will clearly suffice.[4] We assume at once that α, B have been chosen so that $\alpha B^{2(\delta+h)} > A(2(\delta + h))^2$, so the starting point of the induction is guaranteed. Let us assume the inequality to be true up to an $n \geq 2(\delta + h)$. It will suffice to prove it for $m = n + 1$.

For this purpose we proceed to estimate the maximum on the right of (B.2).

In each term of the relevant sum we bound by A each $|a_j|$ with $j \leq 2(\delta + h)$. Taking into account the possible positions of these terms (at most 2^δ in number) and the possible values of the corresponding indices (at most $(2(\delta+h))^\delta$ variations), we see that the sum does not exceed $c_2 A^\delta$ times the sum of the same quantities, but where we restrict the indices to satisfy $2(\delta + h) < j_1, \ldots, j_s \leq n$ and $j_1 + \ldots + j_s \in [n + h - \delta(2\delta + 2h + 1), n + h + 1]$. To bound the new maximum arising in this way we distinguish a few cases depending on s.

If $s = 0$ there is only the 'empty product' term, which we estimate by 1.

If $s = 1$, the sum contains at most $\delta(2\delta + 2h + 1) + 1 \leq c_3$ terms, each bounded by $\alpha B^n/n^2$ by induction.

If $s \geq 2$, one index j_i at least will be larger than $(n - \delta)/s \geq n/2s$. The position of this index may vary at most in s ways, so, isolating this term, we see by induction that the relevant sum is bounded by

$$4\alpha^s \frac{B^{n+h+1}}{n^2} s^3 \sum_{j_1, \ldots, j_{s-1}} \frac{1}{j_1^2 \cdots j_{s-1}^2} \leq c_4 \alpha^2 \frac{B^{n+h+1}}{n^2},$$

because $\sum_{j=1}^{\infty} \frac{1}{j^2}$ is convergent. By (B.2) we have then the inequality

$$|a_{n+1}| \leq 2c_1 + c_1 c_2 A^\delta \max\left(c_3 \alpha B^n/n^2, c_4 \alpha^2 \frac{B^{n+h+1}}{n^2}\right),$$

---

[4] In the kind of proof below the shape of a bound '$|a_m| \leq \alpha B^m$' (though equivalent to the above one for our purposes) does not allow the induction to work.

whence, setting $\beta := \alpha B^h$,

$$\frac{|a_{n+1}|(n+1)^2}{\alpha B^{n+1}} \leq \frac{2c_1(n+1)^2}{\beta B^{n+1-h}}$$
$$+ c_1 c_2 A^\delta \left(c_3(n+1)^2/n^2 B + c_4 \beta(n+1)^2/n^2 \right).$$

To conclude, we first choose a positive $\beta < (8c_1 c_2 c_4 A^\delta)^{-1}$, so that the last term arising on the right is $< 1/2$. We then choose a $B > (8c_1/\beta) + 16c_1 c_2 c_3 A^\delta$ and also $B > A(2(\delta + h))^2/\beta$ to satisfy an earlier requirement; finally we find back $\alpha := \beta/B^h$. Since $n+1-h > 1+(n/2)$, we have $B^{n+1-h} > B(n+1)^2$, and then these inequalities show that the left side of the last displayed inequality must be < 1, concluding the induction step and proving the sought conclusion.   □

Note that (B.1) yields also that the a_n are uniquely determined by $a_0 = 0, a_1, \ldots, a_h$. The recurrence (B.1) further implies that the ring $k[a_1, a_2, \ldots]$ is finitely generated over k, providing in particular an alternative argument for part of Theorem B.1.1. When $h = 0$, i.e. when $f_Y(0, 0) \neq 0$, we find that $y(x)$ itself is uniquely determined (in practice determined by the initial condition $y(0) = 0$). This is a formal analogue of Dini Theorem, and also a special case of Proposition 2.6.3.

Next, we also prove that the values so obtained yield solutions of the equation $f(x, y) = 0$. For this we may and will assume that k is complete with respect to $| \cdot |$.

Theorem B.1.3. *In the above notation, if $|a_m| \leq B_1 B_2^m$ for a $B_2 > 0$ and if $\xi \in k$ is such that $|\xi| < B_2^{-1}$, then the value $y(\xi) \in k$ is such that $f(\xi, y(\xi)) = 0$.*

Proof. We preserve the notation of the previous proof. Observe that $f(x, y_n(x))$ is a polynomial of degree $\leq n \deg_Y f + \deg_X f$, of the shape $\sum_{m \geq n+1} b_{m,n} x^m$, where the coefficients $b_{m,n}$ come from a sum of terms of the form $P(x) y_n(x)^l$, for P a coefficient of f as a polynomial in Y and $l \leq \delta$. Then it is immediately checked that they satisfy

$$|b_{m,n}| \leq c_4 B_1^\delta (m+1)^\delta B_2^m.$$

Now, let $\eta := |\xi| B_2 < 1$. Then the previous estimate yields $|f(\xi, y_n(\xi))| \leq c_5 B_1^\delta \frac{\eta^n}{(1-\eta)^{\delta+1}}$. This tends to 0 as $n \to \infty$; also, $y_n(\xi)$ converges to the sum $y(\xi)$, because $\eta < 1$ and k is complete. Then by continuity we have $f(\xi, y(\xi)) = 0$, as required.   □

The above recurrence arguments easily yield also:

Theorem B.1.4 (Eisenstein Theorem). *Suppose that there is a number field k containing all the a_n. Then there exists a positive integer D such that $D^n a_n$ is an algebraic integer for all n. In particular, there exists a finite set S of DVRs of k such that $a_n \in \mathcal{O}$ for all n and for all $\mathcal{O} \notin S$, so the v-adic radius of convergence is ≥ 1 for all $\mathcal{O} \notin S$.*

Note that by Theorem B.1.1, if f has algebraic coefficients necessarily there exists a number field containing all the a_n.[5]

Proof. We start by proving the penultimate assertion. Notation being as above, let S be the set of DVRs $(\mathcal{O}, \mathcal{P})$ of k such that either (i) $b \in \mathcal{P}$ or $f \notin \mathcal{O}[X, Y]$ or (iii) there exists $j \leq h$ with $a_j \notin \mathcal{O}$.

Note that S is indeed finite: By Corollary 2.5.4, each $u \in k$ lies in $\mathcal{O}$ up to finitely many exceptions $\mathcal{O}$. Taking the union of these finite sets corresponding to the finitely many u's which arise (we take $u = 1/b$, $u =$ some coefficient of f and $u = a_1, \ldots, u = a_h$) we get a finite set containing S. Now, formula (B.1) shows that if $\mathcal{O} \notin S$ and if $a_1, \ldots, a_n \in \mathcal{O}$ for an $n \geq h$, then $a_{n+1} \in \mathcal{O}$. Then by induction it follows that all the a_n lie in $\mathcal{O}$ except possibly if $\mathcal{O} \in S$.

Let now $(\mathcal{O}, \mathcal{P}) \in S$ and endow k with the absolute value $|\cdot|$ corresponding to $\mathcal{P}$, choosing *e.g.* $c = 2$; that is, for $u \in k$ we set $|u| := 2^{-v(u)}$ where v is the order function relative to $\mathcal{P}$. By Theorem B.1.2, there exists a positive B such that $|a_n| \leq B^{n+1}$ for all n. This implies that $v(a_n) \geq -l(n+1)$ for some positive $l = l_{\mathcal{O}}$ and all $n \geq 0$. Let $\mathcal{P} \cap \mathbb{Z}$ be generated by the prime number p, so $v(p) > 0$. Then, plainly there exists $q = q_{\mathcal{O}}$ such that $p^{q_n} a_n \in \mathcal{O}$ for all n: it suffices to take $q > (l+1)/v(p)$.

Finally, take D to be the product of the finitely many prime powers p^q which arise in this way from the $\mathcal{O} \in S$. Then $D^n a_n \in \mathcal{O}$ holds for all n and for all DVRs $\mathcal{O}$ of k, without exceptions. This means that the $D^n a_n$ are algebraic integers (by Application E of Theorem 2.2.1). The assertion on the radius of convergence is clear: If $a_n \in \mathcal{O}$ we have $|a_n| \leq 1$ for the corresponding absolute value. $\qquad\square$

---

[5] In fact, each solution $y(x)$ has automatically coefficients which are algebraic over k. Actually, we are assuming this last fact here, but one can see it either by recalling from Exercise 2.6.2 that each root of f is a Puiseux series over $\bar{k}$, or also directly: by implicit differentiation one finds that all derivatives $y^{(m)}(t) = m! a_m + O(t)$ are algebraic over $k(t)$ and by setting $t = 0$ one finds that a_m is algebraic over k.

Exercise B.1.2.

(i) Use Eisenstein Theorem to prove that neither $\log(1 + z)$ nor $\exp x$ represent algebraic functions.

(ii) Verify directly the Eisenstein condition for rational functions.

(iii) Verify directly the Eisenstein condition for the binomial expansion of $(1 + x)^r, r \in \mathbb{Q}$. Which primes can we find in the denominators of the coefficients?

(iv) Let $r \in \mathbb{Q}$ and consider the binomial series $s(x) = (1 + x)^{\sqrt{r}}$ (for some choice of the square root). Prove that this series is algebraic if and only if r is a perfect square in $\mathbb{Q}$.

(Hint: consider the complex function it represents for small x and then continue it to large real x; equivalently, put $x = \exp(z) - 1$ for small complex z. Alternatively, an equation $f(x, s(x)) = 0$ yields $\sum_{i=1}^{m} c_i (1 + x)^{\mu_i} = 0$, where $\mu_i \in \mathbb{Z} + \sqrt{r}\mathbb{Z}$. Differentiating one obtains an independent relation, and by induction on the number of terms one may conclude.)

Remark B.1.2. The result is still true in positive characteristic, for primes p such that r is a quadratic residue, and the last argument may be adapted to this (after a normalisation to suppose the μ_i not all divisible by p in $\mathbb{Z}_p$). In this case, note that the series makes sense by reduction modulo p of the series in characteristic zero: if r is quadratic residue, then $\sqrt{r} \in \mathbb{Z}_p$ and hence the binomial coefficients lie also in $\mathbb{Z}_p$ (see Exercise 2.6.1(vi)) and may be reduced modulo p. Still, $\sqrt{r}$ is not an integer, which allows the previous argument to work. One may proceed more formally as follows. For $\mu \in \mathbb{Z}_p$ we may define the formal series $s_\mu(x) := (1 + x)^\mu = \sum \binom{\mu}{m} x^m \in \mathbb{Z}_p[[x]]$ and consider its reduction $\tilde{s}_\mu \in \mathbb{F}_p[[x]]$. By reduction one derives the usual formal properties $\tilde{s}_\mu(x)\tilde{s}_\nu(x) = \tilde{s}_{\mu+\nu}(x)$ and $(1 + x)\tilde{s}'_\mu(x) = \mu\tilde{s}_\mu(x)$, where a dash denotes formal differentiation. We have also $\tilde{s}_{p\mu}(x) = \tilde{s}_\mu(x^p)$, for $\mu \in \mathbb{Z}_p$.[6] One may then prove the following: *If $\mu_1, \ldots, \mu_l \in \mathbb{Z}_p$ are distinct, then the $\tilde{s}_{\mu_i}$ are linearly independent over any finite field.* This proceeds as outlined above: if $\sum_{i=1}^{l} c_i \tilde{s}_{\mu_i}(x) = 0$ is a nontrivial relation with minimal length l, then we may assume, on multiplying by $\tilde{s}_{-\mu_1}$, that $\mu_1 = 0$. We may also assume, on changing x with some x^{p^h}, that not all the μ_i are divisible by p. Now differentiation leads to a nontrivial relation with a smaller number of terms, and induction applies. This is

---

[6] Note that, of course, it is not true that $\tilde{s}_\mu$ is generally a series of the shape $(1 + x)^\eta$ for some $\eta \in \mathbb{F}_p$. However one may represent $\tilde{s}_\mu(x)$ in the shape $\prod_{i=0}^{\infty} \tilde{s}_{a_i}(x^{p^i}) = \prod_{i=0}^{\infty}(1 + x^{p^i})^{a_i}$, where $\mu = a_0 + a_1 p + \ldots$, the a_i being integers, $0 \le a_i \le p - 1$.

a kind of 'Lindemann Theorem'[7] for power series in characteristic p; it easily implies that if $1, \mu_1, \ldots, \mu_l$, are linearly independent over $\mathbb{Q}$, the $\tilde{s}_{\mu_i}$ are algebraically independent over any $\mathbb{F}_q(x)$.

As a related **exercise**, prove that for $\mu \in \mathbb{Z}_p \setminus \mathbb{Z}$, the sequence $\left(\binom{\mu}{n}\right)_{n \in \mathbb{N}}$ is not eventually periodic modulo p.

Prove that the Eisenstein condition is satisfied if and only if r is a square modulo all but finitely many prime numbers.

(It may be proved by the Tchebotarev Theorem - see [18] - that this last fact implies that r is in fact a square in $\mathbb{Q}$. An elementary argument, due in essence to Chudnovski, runs as follows: For large N and $M = o(N)$ consider $\prod_{t=1}^{M}\left(\binom{t\sqrt{r}}{N}\binom{-t\sqrt{r}}{N}\right)$. Under the relevant assumption, this is 'almost' an integer, which one can bound by something like cN^{M^2}. But its numerator may be easily shown - pigeon hole principle - to be divisible by every prime in $[N+1, NM/2]$. Hence, choosing e.g. $M \approx \sqrt{N}$, and using Chebishev's well-known elementary prime-number estimates, we conclude that the product must vanish.)

(v) Produce transcendental power series with rational coefficients, which nevertheless satisfy the Eisenstein condition.

(Hint: Use for instance cardinality considerations for existence or consider series with 0 radius of convergence or look for instance at strongly lacunary power series - see for this Exercise (viii) below.)

Though there exist transcendental power series satisfying the Eisenstein condition, for certain classes of functions, satisfying additional assumptions, like being solutions of suitable linear differential equations over $k(x)$, the condition is necessary and sufficient for algebraicity. This kind of results (related to a celebrated conjecture of Grothendieck) are far from trivial already for the binomial expansion for the functions $(1 + x)^\alpha$ with algebraic α. (See Exercise (iv) above and [10] and the related references to work by G.V. Chudnovski, T. Honda, N. Katz and others.)

(vi) Prove that a Puiseux series $y(x) \in k[[x]]$ over a number field k has radius of convergence $= 1$ at almost all valuations, or it is a polynomial.

(Hint: If the radius is > 1 for a DVR of k, the reduction of $y(x)$ modulo $\mathcal{P}$, if defined at all, is a polynomial. Suppose this happens for infinitely many DVRs. Letting $f(x, y(x)) = 0$ be a minimal equation, we then deduce from Ostrowski Theorem=Exercise A.1.1(vii) - one will need a version over a general number field - that $\deg_Y f = 1$. Alternatively, note that if

---

[7] We are alluding here to Lindemann's famous theorem on algebraic independence of values of the exponential function at algebraic arguments, of which the transcendence of π is a very special case.

the solution reduces modulo $\mathcal{P}$ to a polynomial, this polynomial must have bounded degree; hence this can happen for at most finitely many primes, if there are no polynomial solutions. This also provides an alternative argument for Ostrowski Theorem.)

(vii) Let $y(x) = \sum_{m=1}^{\infty} a_m x^m \in k[[x]]$ be a Puiseux series of an algebraic function. Define $Z = Z_y = \{m \in \mathbb{N} : a_m = 0\}$.

Prove that if $y(x) \in k(x)$ then Z is a union of a finite set with a finite set of arithmetic progressions. (See Exercise 2.6.7(xiv)=Theorem of Skolem et al..)

Suppose now that for infinitely many prime numbers l, Z contains a whole arithmetic progression modulo l. Prove that then y is integral over $k[x]$.

(Hint: If Z contains the progression $-a + l\mathbb{Z}$ we have $\sum_{\theta^l = 1} \theta^a x^a y(\theta x) = 0$.

The series $y(\theta x)$ represent algebraic functions ψ_θ, generating altogether a function field L over $k(x)$; let Γ_θ be the set $x(P)$ where P runs over the poles of ψ_θ not lying over the infinite point of $k(x)$. Then $\Gamma_\theta = \theta^{-1}\Gamma_1$, so the identity implies that two distinct elements of Γ_1 have a ratio in the set of l-th roots of unity. If this happens for infinitely many l, Γ_1 must be empty, so the only poles are poles also of x.)

By considering ramification in $k(x, y)/k(x)$ this result may be improved to the conclusion that y is a polynomial; in fact, the relation $\sum_{\theta^l = 1} \theta^a x^a y(\theta x) = 0$ implies that the set of nonzero finite branch points of the extension $k(x, y)/k(x)$ either is empty or contains two distinct points whose ratio is an l-th root of unity. This last fact cannot happen for an infinity of l, so 0 and ∞ are the only branch points. But then $k(x, y)/k(x)$ is totally ramified above 0. However the series expansion of y shows that this is impossible unless the degree is 1; so y is a rational function and by the previous result it must be a polynomial.

(viii) Notation being as in the previous exercise, suppose that Z contains the integers in the interval $[a, b]$ Prove that b/a is bounded dependently only on y (not on a, b), or $y(x)$ is a polynomial. (Use, e.g., (B.1) above, which gives an explicit bound).

In characteristic 0 it may be proved that $b - a$ is bounded if y is not a polynomial.[8] In positive characteristic this need not be true, as shown by the function $y(x) := x + x^p + x^{p^2} + \ldots$, which in characteristic p satisfies $y^p - y + x = 0$.

---

[8] One can use, for instance, a linear differential equation with polynomial coefficients satisfied by the function. It is easy to see that a nontrivial such equation exists, on looking at the dimension of the space spanned by the derivatives over the field $k(x)$

(ix) Let $y(x) = \sum_{m=0}^{\infty} a_m x^m \in \mathbb{C}[[x]]$ be a series converging in, say, the open unit disk and having the unit circle as a natural boundary (for analiticity). Prove that $y(x)$ cannot represent an algebraic function. (Hint: Otherwise $y(x)$ satisfies an irreducible equation $F(x, y(x)) = 0$; by Theorem B.1.2 and analytic continuation theory it may be then continued along any path in $\mathbb{C}$ starting inside the unit disk and remaining in a simply connected domain non containing any point x_0 with $F(x_0, Y)$ having multiple roots or lowered degree.)

Deduce that the *Fredholm series* $y(x) := \sum_{m=0}^{\infty} x^{2^m}$ does not represent an algebraic function. (Putting $x = r\theta$, where $0 < r < 1$, θ is a fixed 2^h-th root of unity and letting $r \to 1$, we see that $y(r\theta) \to \infty$, whence y cannot be continued on any connected open set containing the unit open disk.)

Alternative arguments for this last result come from remark to Exerc. (viii) or directly from consideration of a minimal equation for $y(x)$ and the functional equation $y(x^2) = y(x) - x$. As a related **exercise**, analyse the algebraicity question in positive characteristic.

Appendix C
Discrete valuation rings and Dedekind domains

C.1. Dedekind domains

In this appendix we shall present further descriptions of discrete valuation rings and we shall also briefly discuss Dedekind domains, of which the DVRs are the local version; important examples of them occur with (i) the rings of algebraic integers in a number field and (ii) the ring of elements in a function field K/k in one variable which have poles at most in a prescribed finite set of places. An example of the last type is the integral closure of $k[x]$ in K, when x is transcendental over k.

We recall that a domain is said to be of (Krull) *dimension 1* if every nonzero prime ideal is maximal. (In general, the Krull dimension of a domain is the length of a maximal chain of prime ideals.) We start with a characterization of DVRs:

Theorem C.1.1. *Let A be a domain. Then A is a DVR if and only if it is local, Noetherian, integrally closed and of dimension 1.*

Proof. If A is a DVR, it is a principal domain (hence Noetherian) with a single nonzero prime (hence local and of dimension 1) by definition. In particular it is a VR and so integrally closed (Proposition 2.1.3). This proves half of the statement.

For the other half, let M be the maximal ideal of A and note that since A is Noetherian there exists $\pi \in M$ which is irreducible in A, *i.e.* it cannot be written as the product of two non-units in A (otherwise we could easily construct an infinite ascending chain of principal ideals).

Let us define by recurrence a sequence of elements $a_1, a_2, \ldots \in A \setminus \pi A$ and a corresponding ascending chain of nonzero proper ideals $I_n := \{x \in A : x a_n \in \pi A\}$. We set $a_1 = 1$ and we go on by induction as follows. Suppose a_n defined in such a way that I_n is proper (which means that $a_n \notin \pi A$). Then, if $I_n = M$ the procedure stops. Otherwise, I_n is not prime (because it is nonzero and A has dimension 1) so there exist $r, s \in A \setminus I_n$ with $rs \in I_n$. We then define $a_{n+1} := r a_n$.

Since $r \notin I_n$ we have $a_{n+1} \notin \pi A$ as required, which also implies that I_{n+1} is proper. Moreover, the ideal $J := I_n + sA$ contains properly I_n and certainly I_{n+1} contains J and so contains I_n, properly because $s \notin I_n$.

Since A is Noetherian, the procedure will eventually stop,which means that $I_l = M$ for some l. Define then $a := a_l$; we have $aM = \pi B$ for some ideal $B \subset A$. Suppose that B is a proper ideal; then $B \subset M$ and, putting $x := a/\pi$ (an element of the fraction field K of A), we have $xM \subset M$. But M is finitely generated (since A is Noetherian), and then x is integral over A and hence lies in A because A is integrally closed by assumption. This is however a contradiction with the fact that $a = a_l \notin \pi A$, and therefore $B = A$. Hence there exists $m \in M$ with $am = \pi$. But π is irreducible, so this implies that $a \in A^*$ and therefore, since $aM = \pi A$, we deduce that π divides every element of M. Hence $M = \pi A$ is principal.

The rest follows easily. First, let $J := \cap_{n=1}^{\infty} \pi^n A$. Then J is an ideal satisfying $J = MJ$. Since J is finitely generated we must have $I = 0$ by Nakayama's lemma.

Then for a nonzero element $x \in A$ we may define $v(x) := \max\{n \in \mathbb{N} : x \in \pi^n A\}$ as a function with values in $\mathbb{N}$. Clearly $v(x) = 0$ if and only if $x \in A^*$. Let then I be a nonzero ideal of A and let $m := \min\{v(x) : x \in I\}$. It is clear that $I = \pi^m A$, so A is principal, completing the proof. $\qquad\square$

We now turn to Dedekind domains:

Definition C.1.1. A *Dedekind domain* (abbreviated DD) is a Noetherian integrally closed domain of dimension 1.

Note that by Theorem C.1.1 the local Dedekind domains are precisely the discrete valuation rings.

Proposition C.1.2. *Let A be a DD. Then A is the intersection of a set of DVRs of its quotient field K. Also, each localization of A at a prime ideal is a DVR of K.*

Proof. By Application E of Theorem 2.2.1, the integral closure of A in K is the intersection of the VRs of K containing A; since A is integrally closed, this intersection is A itself. Let $(\mathcal{O}, \mathcal{P})$ be some VR of K containing A. The intersection $M = M_{\mathcal{O}} := \mathcal{P} \cap A$ cannot be 0, for otherwise $A \setminus \{0\}$ would be contained in $\mathcal{O}^*$, hence $\mathcal{O}$ would equal K, a contradiction. Therefore M is a maximal ideal of A; also, A_M is a local integrally closed Noetherian domain of dimension 1, hence a DVR by

Theorem C.1.1 Note that $A_M \subset \mathcal{O}$ and so $A_M = \mathcal{O}$ by Proposition 2.4.1 This proves that $\mathcal{O}$ is a DVR of K; hence A is the intersection of a set of DVRs of K.

The rest of the statement also follows as a byproduct of this argument.   □

Example C.1.1. The ring of integers in a number field id a DD, and also the integral closure of $k[x]$ in a function field of one variable K/k, where x is transcendental over k and $K/k(x)$ is separable. In fact, in each case such a ring A is integrally closed (by transitivity of integral property) and Noetherian, because it is a finite module over $\mathbb{Z}$ or $k[x]$ respectively (this follows from Theorem 2.5.2). Also, A has dimension 1. In fact, let P be a nonzero prime ideal of A. Then P is necessarily maximal, because the residue ring A/P is integral over the field $\mathbb{Z}/(P \cap \mathbb{Z})$ or $k[x]/(P \cap k[x])$; hence A/P is also a field.

The terminology *dimension 1* comes from the function-field case, when the set of points corresponds to a *curve*.

C.2. Factorization

In the sequel let A be a DD and let P run through the set Σ_A of its maximal ideals.

Proposition C.2.1. *Let I be a nonzero ideal of A. Then the set of prime ideals of A which contain I is finite.*

Proof. We argue by contradiction. We order by inclusion the set of ideals I violating the conclusion. If this set is not empty, it has a maximal element I_0, since A is Noetherian. If I_0 is prime, it is maximal, since A has dimension 1, so the conclusion would be true. Hence I_0 is not prime, so there exist $x_1, x_2 \in A \setminus I_0$ such that $x_1 x_2 \in I_0$. Then both $I_j := I_0 + x_j A, j = 1, 2$, are larger than I_0 and so satisfy the conclusion; also, clearly $I_1 I_2 \subset I_0$. Let now P be a prime ideal containing I_0. Then P contains $I_1 I_2$, whence it contains either I_1 or I_2 (otherwise pick $a_j \in P \setminus I_j$ to obtain $a_1 a_2 \in I_1 I_2 \setminus P$). Hence P is in a finite set, a contradiction which concludes the argument.   □

Theorem C.2.2. *In a DD every nonzero ideal can be written uniquely as a product of prime ideals.*

Proof. Let $I \subset A$ be a nonzero ideal and let us consider, for each $P \in \Sigma_A$, its extension I_P to the localized ring $A_P := \{a/b : a \in A, b \in A \setminus P\}$; by definition, $I_P = I A_P$. Since A_P is a DVR, we have $I_P = (P_P)^{a_P}$

where $a_P \geq 0$ is an integer. Note that a_P can be positive only if $P \supset I$, and by the last proposition this can happen only for finitely many P.

Note also that $P^a = (P_P)^a \cap A$: indeed, if $x \in (P_P)^a \cap A$ there exists $y \in A \setminus P$ such that $yx \in P^a$; now, since P is maximal there exists $z \in A$ with $yz \equiv 1 \pmod{P}$, whence, raising $yz - 1$ to the power a we see that there exists z' with $yz' \equiv 1 \pmod{P^a}$. Hence $x \equiv (y'y)x \equiv 0 \pmod{P^a}$, as required.

Let us consider the ideal $J := \cap_P P^{a_P}$. Since $I = I_P \cap A \subset (P_P)^{a_P} \cap A = P^{a_P}$ (by the previous argument), this ideal contains I. Moreover, the ideals in Σ are pairwise comaximal, whence the same is true of positive powers of them. Hence (by a standard argument) $J = \prod_P P^{a_P}$ and $A/J \cong \prod_P (A/P^{a_P})$.

Finally, observe that for every $x \in J$ and every $P \in \Sigma$ there exists $c_P \in A \setminus P$ with $c_P x \in I$: in fact, $x \in P^{a_P} \subset I_P$. Put $B := \sum_P c_P A$; then $Bx \subset I$. On the other hand B is not contained in any maximal ideal of A, whence it is A. Hence $x \in I$.

All of this proves that $J = I$, so I is indeed a product of maximal ideals.

To prove uniqueness, let $I = \prod_P P^{b_P}$ for integer exponents $b_P \geq 0$ (almost all equal to 0). Then, for $Q \in \Sigma_A$, $I_Q = \prod_P (P_Q)^{b_P} = Q^{b_Q}$, proving that $b_Q = a_Q$ is uniquely determined, as asserted. $\square$

Remark C.2.1. Note that the theorem implies that: *A prime ideal P divides the ideal I* (i.e. $I = PQ$ *for some ideal Q) if and only if P contains I*. One half of this statement is clear (and valid in any ring). As to the converse, suppose that P contains I (supposed $\neq 0$). Then, factoring $I = P_1 \cdots P_s$ as a product of maximal ideals P_i, we see that P contains some P_i (otherwise pick $a_i \in P_i \setminus P$ and form $a_1 \cdots a_s$ to get a contradiction). Since P_i is maximal, we have $P = P_i$, as required. (A slight variation of the argument shows that the statement holds also for non-prime P.)

We conclude with some comments on Unique Factorization. Of course, this occurs when A is a Principal Ideal Domain; this is certainly not always the case, as is well-known. At the moment we only note the following:

Proposition C.2.3. *A Dedekind domain A has unique factorization if and only if A is principal.*

Proof. We have already recalled the standard easy fact that a principal-ideal domain has unique factorization.

For the converse, suppose that A is a UFD and that P is a non-principal maximal ideal. Pick then $x \in P \setminus P^2$; then the factorization of Ax (we are using Thm. A3.2) takes the form PQ where the ideal Q is not divisible by P. We can factor x into irreducible elements and hence we may assume that x is in fact irreducible. Now, Q is not contained in P (Remark to Thm. A3.2) and we may pick $y \in Q \setminus P$. Then $Ay = QR$ where P does not divide R. Since P is maximal and $Q \neq P$, A (if $Q = A$, P would be principal), certainly P is not contained in Q, so there exists $z \in P \setminus Q$, whence $Az = PS$ for some ideal S. This implies that the ideal Ayz is divisible by the ideal Ax: $Ayz = (Ax)J$. Hence $yz = xw$ with $w \in J$. Since x is irreducible, it has to divide either y or z. However, it cannot divide y, because $y \notin P$ and it cannot divide z, because $z \notin Q$. This contradiction proves the statement.   □

C.3. Notes to appendices

Appendix A
Several proofs for the Nullstellensatz are known, more or less direct. See [2] for a simple inductive proof due to Zariski, based on a few facts on integral elements, and see [17] or [19] for proofs by elimination, using resultants. More recently, 'effective' versions of the theorem have been studied by several authors; proofs have been found which give (good) bounds for the exponent and degrees which arise in expressing a power of a given element of the radical $\sqrt{J}$ in terms of generators for J.

The 'deep fact' alluded to in Exercise A.1.1 is the *Lang-Weil* Theorem, which estimates the number of points of a variety over a finite field; in particular, it implies that an absolutely irreducible variety over $\mathbb{Q}$ has points modulo p for all but finitely many primes p. This may be derived from Weil's 'Riemann Hypothesis' for curves over finite fields. (See [12].)

Ostrowski Theorem (appearing as Exercise A.1.1(viii)) may also be proved by elimination theory (see [31]).

Appendix B
In the complex case convergence may be also proved by complex integration, proving local analiticity of solutions. For other approaches to convergence (by Cauchy's 'method of majorants') see [1, 10] or [28], where general linear differential equations over $k(x)$ are also considered. In [10] and especially in B. Dwork, A.J. van der Poorten, The Eisenstein constant, *Duke Math. J.*, 1992, one can also find sophisticated investigations of the radii of convergence, leading to sharp quantitative versions of Eisenstein Theorem. (We remark that the complex case is easier for this,

because of analytic continuation: now the radius of convergence is the distance to the nearest singularity. In the ultrametric case analytic continuation generally fails, because of total disconnectedness, leading to substantial difficulties.) For other elementary proofs of Eisenstein Theorem see [15] or [23].

Appendix C
The argument for Theorem C.1.1 follows [24], with a slight variation.

For the rest, our treatment is more or less standard. See also [6, 18, 24] or [21].

References

[1] E. ARTIN, "Algebraic Numbers and Algebraic Functions", Nelson, 1968.

[2] M. ATIYAH and I. MACDONALD, "Commutative Algebra", Addison Wesley, 1969.

[3] E. BOMBIERI and W. GUBLER, "Heights in Diophantine Geometry", New Math. Monographs n. 4, Cambridge Univ. Press, 2006.

[4] M. BAKER and R. RUMELY, "Potential Theory and Dynamics on the Berkovich Projective Line", Mathematical Surveys and Monographs, 159, American Mathematical Society, Providence, RI, 2010.

[5] I. BOREVITCH and I. SHAFAREVITCH, "Théorie des nombres", Gauthier Villars, 1967.

[6] J. W. S. CASSELS and A. FRÖLICH (eds.), "Algebraic Number Theory", 1967.

[7] C. CHEVALLEY, "Introduction to the Theory of Algebraic Functions in One Variable", AMS Math. Monographs, 6, 1951.

[8] P. CORVAJA and U. ZANNIER, *Arithmetic in infinite extensions of function fields*, Boll. U.M.I., 1997.

[9] V. DANILOV, *Algebraic varieties and schemes*, In: "Algebraic Geometry", I, Encyclopoedia of Math.Sciences, I.R. Shafarevic Ed., Springer-Verlag, 1994.

[10] B. DWORK, G. GEROTTO and F. SULLIVAN, "An Introduction to G-functions", Annals of Math. Studies, Princeton Univ. Press, 1992.

[11] O. FORSTER, "Riemann Surfaces", Springer Verlag, GTM 81, 1981.

[12] M. FRIED and M. JARDEN, "Field Arithmetic", Springer-Verlag.

[13] W. FULTON, "Algebraic Topology - A First Course", Springer-Verlag, GTM 153, 1995.

[14] R. HARTSHORNE, "Algebraic Geometry", Springer-Verlag.

[15] M. HINDRY and J. H. SILVERMAN, "Diophantine Geometry", Springer Verlag, 2004.

[16] S. LANG, "Introduction to the Theory of Algebraic and Abelian Functions", Springer Verlag, GTM 89, 1972.

[17] S. LANG, "Algebra", Springer-Verlag.

[18] S. LANG, "Algebraic Number Theory", Addison Wesley.

[19] M. MANETTI, "Corso Introduttivo alla Geometria Algebrica", Appunti, Scuola Normale Superiore, Pisa, 1998.

[20] V. MANTOVA and U. ZANNIER, *Artin-Whaples approximations of bounded degree on algebraic varieties*, Proceedings of the American Mathematical Society, to appear.

[21] W. NARKIEVICZ, *Elementary and Analytic Theory of Algebraic Numbers*, Polish Scientific Publishers, Warszawa, 1990.

[22] M. OJANGUREN, "The Witt Group and the Theorem of Lüroth", ETS Ed., Pisa.

[23] G. POLYA and G. SZEGO, "Problems and Theorems in Analysis II", Springer Verlag, 1976.

[24] J-P. SERRE, "Corps Locaux", Hermann, Paris.

[25] J-P. SERRE, "Algebraic Groups and Class Fields", Springer-Verlag, GTM 117, 1988.

[26] J-P. SERRE, "Course d'arithmétique", Presses Universitaires de France, 1970.

[27] J-P. SERRE, "Topics in Galois Theory", Jones and Bartlett, Boston, 1992.

[28] J-P. SERRE, "Lie Algebras and Lie groups", Benjamin, 1965.

[29] J-P. SERRE, "Lectures on the Mordell-Weil Theorem", Vieweg, 2000.

[30] I. R. SHAFAREVIC, "Algebraic Geometry", Springer-Verlag.

[31] A. SCHINZEL, *Polynomials with special regard to reducibility*, Encyclopedia of Mathematics and its applications, **77**, Cambridge Univ. Press, 2000.

[32] V. V. SHOKUROV, "Riemann Suerfaces and Algebraic Curves, in Algebraic Geometry", I, Encyclopoedia of Math.Sciences, I.R. Shafarevic Ed., Springer-Verlag, 1994.

[33] C. L. SIEGEL, "Topics in Complex Function Theory", Vol. I, Wiley, 1969.

[34] J. H. SILVERMAN, "The Arithmetic of Elliptic Curves", Springer-Verlag, GTM 106, 1986.

[35] H. VOLKLEIN, "Groups as Galois Groups", Cambridge UNiv. Press, 1996.

[36] R. J. WALKER, "Algebraic curves", Springer-Verlag, 1978.

Index

LECTURE NOTES

This series publishes polished notes dealing with topics of current research and originating from lectures and seminars held at the Scuola Normale Superiore in Pisa.

Published volumes

1. M. TOSI, P. VIGNOLO, *Statistical Mechanics and the Physics of Fluids*, 2005 (second edition). ISBN 978-88-7642-144-0
2. M. GIAQUINTA, L. MARTINAZZI, *An Introduction to the Regularity Theory for Elliptic Systems, Harmonic Maps and Minimal Graphs*, 2005. ISBN 978-88-7642-168-8
3. G. DELLA SALA, A. SARACCO, A. SIMIONIUC, G. TOMASSINI, *Lectures on Complex Analysis and Analytic Geometry*, 2006. ISBN 978-88-7642-199-8
4. M. POLINI, M. TOSI, *Many-Body Physics in Condensed Matter Systems*, 2006. ISBN 978-88-7642-192-0

   P. AZZURRI, *Problemi di Meccanica*, 2007. ISBN 978-88-7642-223-2
5. R. BARBIERI, *Lectures on the ElectroWeak Interactions*, 2007. ISBN 978-88-7642-311-6
6. G. DA PRATO, *Introduction to Stochastic Analysis and Malliavin Calculus*, 2007. ISBN 978-88-7642-313-0

   P. AZZURRI, *Problemi di meccanica*, 2008 (second edition). ISBN 978-88-7642-317-8

   A. C. G. MENNUCCI, S. K. MITTER, *Probabilità e informazione*, 2008 (second edition). ISBN 978-88-7642-324-6
7. G. DA PRATO, *Introduction to Stochastic Analysis and Malliavin Calculus*, 2008 (second edition). ISBN 978-88-7642-337-6
8. U. ZANNIER, *Lecture Notes on Diophantine Analysis*, 2009. ISBN 978-88-7642-341-3
9. A. LUNARDI, *Interpolation Theory*, 2009 (second edition). ISBN 978-88-7642-342-0

10. L. AMBROSIO, G. DA PRATO, A. MENNUCCI, *Introduction to Measure Theory and Integration*, 2012.
    ISBN 978-88-7642-385-7, e-ISBN: 978-88-7642-386-4

11. M. GIAQUINTA, L. MARTINAZZI, *An Introduction to the Regularity Theory for Elliptic Systems, Harmonic Maps and Minimal Graphs*, 2012 (second edition). ISBN 978-88-7642-442-7, e-ISBN: 978-88-7642-443-4
    G. PRADISI, *Lezioni di metodi matematici della fisica*, 2012.
    ISBN: 978-88-7642-441-0

12. G. BELLETTINI, *Lecture Notes on Mean Curvature Flow, Barriers and Singular Perturbations*, 2013.
    ISBN 978-88-7642-428-1, e-ISBN: 978-88-7642-429-8

13. G. DA PRATO, *Introduction to Stochastic Analysis and Malliavin Calculus*, 2014. ISBN 978-88-7642-497-7, e-ISBN: 978-88-7642-499-1

14. R. SCOGNAMILLO, U. ZANNIER, *Introductory Notes on Valuation Rings and Function Fields in One Variable*, 2014. ISBN 978-88-7642-500-4, e-ISBN: 978-88-7642-501-1

Volumes published earlier

G. DA PRATO, *Introduction to Differential Stochastic Equations*, 1995 (second edition 1998). ISBN 978-88-7642-259-1

L. AMBROSIO, *Corso introduttivo alla Teoria Geometrica della Misura ed alle Superfici Minime*, 1996 (reprint 2000).

E. VESENTINI, *Introduction to Continuous Semigroups*, 1996 (second edition 2002). ISBN 978-88-7642-258-4

C. PETRONIO, *A Theorem of Eliashberg and Thurston on Foliations and Contact Structures*, 1997. ISBN 978-88-7642-286-7

Quantum cohomology at the Mittag-Leffler Institute, a cura di Paolo Aluffi, 1998. ISBN 978-88-7642-257-7

G. BINI, C. DE CONCINI, M. POLITO, C. PROCESI, *On the Work of Givental Relative to Mirror Symmetry*, 1998. ISBN 978-88-7642-240-9

H. PHAM, *Imperfections de Marchés et Méthodes d'Evaluation et Couverture d'Options*, 1998. ISBN 978-88-7642-291-1

H. CLEMENS, *Introduction to Hodge Theory*, 1998. ISBN 978-88-7642-268-3

Seminari di Geometria Algebrica 1998-1999, 1999.

A. LUNARDI, *Interpolation Theory*, 1999. ISBN 978-88-7642-296-6

R. SCOGNAMILLO, *Rappresentazioni dei gruppi finiti e loro caratteri*, 1999.

S. RODRIGUEZ, *Symmetry in Physics*, 1999. ISBN 978-88-7642-254-6

F. STROCCHI, *Symmetry Breaking in Classical Systems*, 1999 (2000).
ISBN 978-88-7642-262-1

L. AMBROSIO, P. TILLI, *Selected Topics on "Analysis in Metric Spaces"*, 2000. ISBN 978-88-7642-265-2

A. C. G. MENNUCCI, S. K. MITTER, *Probabilità ed Informazione*, 2000.

S. V. BULANOV, *Lectures on Nonlinear Physics*, 2000 (2001).
ISBN 978-88-7642-267-6

Lectures on Analysis in Metric Spaces, a cura di Luigi Ambrosio e Francesco Serra Cassano, 2000 (2001). ISBN 978-88-7642-255-3

L. CIOTTI, *Lectures Notes on Stellar Dynamics*, 2000 (2001).
ISBN 978-88-7642-266-9

S. RODRIGUEZ, *The Scattering of Light by Matter*, 2001.
ISBN 978-88-7642-298-0

G. DA PRATO, *An Introduction to Infinite Dimensional Analysis*, 2001.
ISBN 978-88-7642-309-3

S. SUCCI, *An Introduction to Computational Physics*: – Part I: *Grid Methods*, 2002. ISBN 978-88-7642-263-8

D. BUCUR, G. BUTTAZZO, *Variational Methods in Some Shape Optimization Problems*, 2002. ISBN 978-88-7642-297-3

A. MINGUZZI, M. TOSI, *Introduction to the Theory of Many-Body Systems*, 2002.

S. SUCCI, *An Introduction to Computational Physics*: – Part II: *Particle Methods*, 2003. ISBN 978-88-7642-264-5

A. MINGUZZI, S. SUCCI, F. TOSCHI, M. TOSI, P. VIGNOLO, *Numerical Methods for Atomic Quantum Gases*, 2004. ISBN 978-88-7642-130-0